SEARCH:

A Research Guide for Science Fairs and Independent Study

Connie Wolfe, M.A., Education

"Using creative thinking skills as tools for investigation and research helps students apply what they know and are learning to their own area of interest."

Joyce E. Juntune
Executive Director
National Association for Gifted Children

ISBN 0-913705-30-6

Book and Cover Design: Kathleen Koopman
Illustrations: Rodney M. Wolfe
Production coordination and editorial assistance: Kathleen Koopman

ACKNOWLEDGEMENTS

The author wishes to thank many people for their suggestions and help with the compilation of the book **SEARCH.** Many thanks to Joey Tanner for her encouragement and guidance throughout the editing process. And for the many students who offered helpful comments, in particular Rob Knowles and Sophia Pizzo. And a special thanks to my husband, Bill and my sons, Leonard and Rodney, for their patience and support.

CONTENTS

Changingness

"We are, in my view, faced with an entirely new situation in education where the goal of education, if we are to survive, is the facilitation of change and learning. The only man who is educated is the man who has learned how to learn; the man who has learned how to adapt and change; the man who has realized that no knowledge is secure, that only the process of seeking knowledge gives a basis for security; changingness, a reliance on process rather than upon static knowledge, is the only thing that makes any sense as a goal for education in the modern world."

Carl Rogers
Freedom to Learn for the 80's

HOW SEARCH EVOLVED

Students in my classes, all the way from 4th through 9th grades, needed a framework to explore their ideas for science projects and independent study. The students seemed frustrated and unsure of themselves, and their projects lacked quality, in-depth questioning and investigative thinking. I discovered there were few resources available to guide students through a project, or to encourage the use and application of problem solving, and creative and critical thinking skills.

With my own sons, I had experienced a similar situation; they needed a guide to allow them to really analyze their work, create, and evaluate what they did in their Science Fair projects. As a child, I remembered my father, Dean Lutz, being my guide through many science and independent study projects that I would not have attempted without his facilitation. He encouraged the "I wonders," inspired me to deepen my knowledge, and use my imagination.

I was finishing my graduate work when I designed strategies enabling students to utilize in-depth thinking while doing experimental or non-experimental research. I felt that if students had a "map" to use, along with a teacher-facilitator, they could produce excellent research and projects. With my background as a science teacher and Science Fair judge, as well as teacher of the gifted, I set about creating **SEARCH.**

To show that students at many levels could do intense science research, I had the first booklet of **SEARCH** classroom-tested with students in grades 4-9, with high school students in science classes, and with those doing independent study. Students began to enjoy research. Self-esteem was high as projects were completed. New interest areas and many career possibilities were found. Students were solving real problems and presenting creative solutions to an appropriate audience where their work was shared and discussed.

Two years later, students who used the guide won wide recognition for their efforts. Many won science fairs and other awards.

The original **SEARCH** was given feedback by the students who used it. Teachers gave other suggestions to add to the guide. My husband, a research and development engineer, provided ideas and dialogued with me over many parts of the book. It is through the efforts of all those who were involved in **SEARCH** that it has become a worthwhile resource guide.

"The mere formation of a problem is far more often essential than its solution, which may be merely a matter of mathematical or experimental skill. To raise new questions, new possibilities, to regard old problems from a new angle requires creative imagination and marks real advances in science."

Albert Einstein

INTRODUCTION

SEARCH is for the Gifted, Talented, or Creative Student, in grades 4 through 9, who wants to do a science project, enter a competitive fair, or for a student who is interested in pursuing an independent study in any subject area.

Each year as part of existing enrichment programs, in programs that offer differentiated curriculum, or through competitive Science Fairs, students present their science and independent study projects for evaluation to teachers, judges and the general public. Students can work in the categories of science: biology, physics, chemistry, computers, the environment and ecology. Some fairs also include engineering and energy as alternative categories. Possible categories of independent study include: social studies (religion, geography, current affairs), local issues (history, culture, politics) and related environmental studies (architecture, ecology, development), and many more.

The first part of **SEARCH: A Research Guide for Science Fairs and Independent Study** is a teacher's guide. It gives directions, guidelines, suggestions and tips for assisting students in the research process.

The second half is the student's section: the guide for the student working independently towards a project or product, developing research skills, problem solving, using higher-level thinking and creativity skills. Through the guidelines in this book, students will be on their way to creating a successful science project or independent study project.

PART I

FOR TEACHERS

SEARCH:
A Research Guide to Science Fairs and Independent Study

The value of self-directed research for the classroom is time honored both in the schools and in the literature. More recently independent study and self-directed research have become the backbone of gifted education and as such have taken on a new and important direction. With this new direction, care is taken that the learner select a topic in the "passion area." (Betts, Autonomous Learning Model, 1985) The student is now urged to become a "practicing professional" (Renzulli, The Enrichment Triad Model, 1977) and to produce a product of interest in the subject field rather than the typical "report to the class."

Students now construct real investigations that end, perhaps, in the printing of their conclusions in a scientific or historical publication. They might research and create an answer to a traffic problem in their community. They could write, choreograph, and produce a musical for a school event. As a group of concerned students, described by Renzulli, they might collect and study the daily trash and paper waste at their school. Then the students could put together a proposal to have the trash weighed and recycled. The money earned would purchase needed equipment for the school.

A young age does not limit the ability to contribute to society. Louis Braille, a blind French student, conceived the idea of an alphabet for the blind when he was ten years old. By the time he was fifteen he had worked out all the details. Examples like these may help your students consider their own abilities and look for ways to make a contribution to others.

The purpose of SEARCH:

1. To assist the student in planning a science project or independent study using research skills in
    a. the acquisition of information
    b. the analysis of data
    c. the concise reporting of information.

2. To communicate to students how to conduct original research through the experi-

mental method. This includes experimental factors and control factors, and collection and recording of the data results.

3. To provide students with a framework for doing a science project or independent study.

4. To encourage students to investigate and research an area of interest, utilizing creative thinking skills, problem solving and critical thinking skills. This includes such skills as application, analysis, synthesis, and evaluation levels of thinking.

5. To encourage students to pursue a particular career interest.

The common question from the student is "WHERE DO I BEGIN?" **SEARCH** shows students how to organize the task of researching and designing a science project or independent study. **SEARCH** supplements the language arts curriculum with a guide to use: the card catalog, tables of contents, indexes, glossaries; how to outline, summarize and draw conclusions. **SEARCH** supplements the math curriculum, where students are taught such skills as measurement, graphing, statistics, probability, and tables, by showing how to acquire information about the student's topic and how to collect and analyze data. **SEARCH** guides the student in product development: how to put the information in a report format or other product, such as a model, a video, or a speech. With this structure for research students achieve their goals easier and their projects turn out to be quality products.

SEARCH provides all of the above by supplying a fully reproducible Student Section that features the **SEARCH** roadmap, a graphic step-by-step guide to creating a research-based independent study or science fair project. This is the unique feature of this book that can be fully utilized by the teacher in a variety of ways. Suggestions, information and forms for your use are at the end of this chapter. You'll find evaluation forms, worksheets and a letter for parents, among others. Also included are classroom management tips, suggestions for arranging your classroom, how to set up a learning center, and tokens to offer the students as they complete each step of this process. The unit is designed for grades 4-9, and can be easily adapted to the high school level. By using the suggestions in this section, the unit can be tailored to the specific needs of your classroom. Allow 8-10 weeks for completion of the unit. If at all possible, obtain all necessary information about upcoming science fairs well in advance. Allow for mailing time for obtaining the many resources available by mail.

The independent study or science product should be a means of communication, i.e., taking new or researched ideas and reconceptualizing them. Students need to become familiar with existing technologies to assist them in developing and communicating the purpose of their products.

Since independent study and science projects are intended for an audience (such as judges at a Science Fair), it is important for students to know how the projects will be evaluated; these criteria will be the basis for how the projects and experiments will be designed and implemented.

What is Independent Study?

The term "Independent Study" means a learning situation within the school day which allows a student to develop personal competencies through experiences as an individual but with interaction with others when needed. It is an in-depth investigation in an area that the student is interested in and wants to know more about. It is characterized by freedom from constant supervision. Students read, write, contemplate, listen to records and tapes, view, record, memorize, create, build, practice, exercise, experiment, examine, analyze, evaluate, investigate, question, discover, and converse and probe the "I wonders."

Independent study **emphasizes** the individual's role in learning. It implies that students possess potentialities for self-initiative, self-discipline, resourcefulness, productivity, and self-evaluation.

Independent Study allows students to:

- Perceive worthwhile things to do
- Personalize learning
- Exercise self-discipline
- Employ their own style of learning
- Make use of human resources
- Make use of material resources
- Solve real problems
- Produce results
- Strive for improvement
- Experience the fact that science is enjoyable and fun

The teacher's role in Independent Study requires the teacher:

- To be open-minded
- To be willing to adapt to new situations rather quickly
- To be more of a counselor and consultant than a traditional teacher
- To be a facilitator who helps guide the student
- To be willing to accept that learning takes place in several different ways
- To be willing to learn along with the students

The teacher therefore takes on a supportive, nurturing role, as an ombudsman who brings out the affective side. The teacher becomes a coach to help with self-esteem, career and counseling.

Thus the role of teacher changes:

From:	**To:**
A fountain of knowledge in subject areas	A consultant
A director of learning	A resource person
A pacer of learning	A manager of learning resources
An enforcer of coverage	An assistant in student self-evaluation
A sage on the stage	A guide by the side

Independent study will provide your students the opportunity to develop responsibility, to learn how to learn and apply what they know. These skills cannot be taught as well in the traditional sense of teaching, but they can be learned in an atmosphere where mistakes are tolerated, even expected, and where professional assistance is available.

—adapted from material accredited to Dr. William M. Griffith

GOALS AND EXPECTATIONS

Science Fair Project: Scope and Sequence

I. The student will select the topic of investigation by

   A. Brainstorming and webbing.

   B. Writing a question about the research topic and narrowing the focus.

   C. Writing a hypothesis in the form of a statement about the question that can be proved or disproved.

   D. Predicting the results.

II. The student will acquire the information needed by

   A. Utilizing computer searches and retrieval systems (such as microfiche and data bases) to find information about the topic.

   B. Writing notecards in the format suggested for notetaking on page 59, checking author's authenticity, source of data, and copyright.

   C. Obtaining information from the Card Catalog, Reader's Guide, vertical files, and science indexes.

   D. Interviewing, making surveys or writing to agencies to obtain data on a topic.

III. The student will design the science experiment or method of study by

   A. Learning the experimental method or the survey method of research.

   B. Understanding and using the experimental factor, control factor, and variable factor to prove or disprove the hypothesis.

   C. Collecting data by means of questionnaires, observation, studying charts and graphs, or writing to authoritative sources; students must keep a log of daily tests and observations.

   D. Analyzing, by categorizing data, comparing and observing data, recording data, and testing the hypothesis especially through the use of computer data bases.

   E. Reporting the data and results by making charts, tables, graphs, diagrams and lists showing measurements of data; using computer printouts from data bases to report variable factors in relation to one another.

IV. The student will interpret information by
   A. Observing carefully the data and what has happened.
   B. Reporting the findings of the data collected.
   C. Using graphs, charts or tables to show measurements.
   D. Drawing conclusions on the findings that are related to the hypothesis and previous research findings.

V. The student will report, create the product and display by
   A. Writing a concise science report with title page, abstract, table of contents, purpose of project, research done on this topic in the past, question or problem, hypothesis (possible solution to the problem), experiment explained, results (data collected and explained), conclusions and summary, and bibliography.
   B. Making a visual display with poster and/or a model to explain the science project.
   C. Completing a product such as those from the Product List (see p. 45).
   D. Entering the science project in a Science Fair or presenting the project to an appropriate audience for evaluation.

Independent Study: Scope and Sequence

I. The student will select the topic by
   A. Brainstorming and webbing.
   B. Writing a question about the research topic and narrowing the focus.
   C. Selecting the right research method, such as case history, historical study, descriptive study, survey method or experimental research method.

II. The student will acquire the information by
   A. Utilizing computer searches and retrieval systems (such as microfiche and data bases) to find information about the topic.
   B. Writing notecards in the format suggested for notetaking; checking author's authenticity, source of data, and copyright.
   C. Obtaining information from the Card Catalog, Reader's Guide, vertical files, and other indexes.

D. Interviewing, making surveys or writing to agencies to obtain data on a topic.

III. The student will develop the research process by
   A. Including content, process and product.
   B. Predicting the results.
   C. Collecting data by means of observation, charts, graphs and logs.
   D. Analyzing data by categorizing, comparing and testing the hypothesis.
   E. Making charts and graphs showing measurements of data, where appropriate.

IV. The student will interpret information by
   A. Observing carefully the data and factors collected.
   B. Reporting the findings of the data collected.
   C. Using graphs, charts, logs or tables to show measurements.
   D. Drawing conclusions on the findings that are related to their own predictions and observations as well as to other research findings.

V. The student will create a report, product and display by
   A. Presenting the independent study in written form with: title page, abstract, table of contents, purpose of independent study, body of report (research found and student's independent study data explained), conclusion and summary, and bibliography.
   B. Making a visual display poster and/or a model to explain the independent study.
   C. Creating a product (see the Product List, p. 45).
   D. Entering the independent study in a competition or presenting it to the appropriate audience for evaluation.

TEACHING STRATEGIES

Provide each student with their own copy of Part II of **SEARCH**. Each student will review the introductory material, complete a few practice worksheets, and then begin the steps as outlined. **SEARCH** provides the students with:

A. A specially designed road map that encourages and motivates students to work through the steps of research toward an end product and enter their project in a Science Fair or for evaluation to a certifier.

B. Worksheets structured for specific tasks, such as finding area of interest, brainstorming or writing the hypothesis.

C. A certificate at the end of the book rewarding students for the completed task; a positive reinforcement for their effort.

D. Many examples for the students to follow when completing a task in the book.

E. Lists of indexes and sources to start beginning students on the research task.

F. And finally, self-evaluation sheets to help students look at their own work objectively and think about possible changes or improvements.

A good way to integrate **SEARCH** into your classroom is to create a Learning Center to facilitate learning research methods. This center would provide:

- A poster of the map guide to **SEARCH.**
- The game called "**SEARCH LINGO**" at the end of this section.
- Samples of other student science projects or independent studies for motivation. These could be photographs or the actual projects.
- Notecards, with a sample for students to follow.
- A bulletin board to illustrate student work and keep classmates posted on each others' progress.
- Additional research materials/resources such as:
  The book Thousands of Science Projects may help you decide on a topic. It can be ordered for $3.00 from Science Service, 1719 N. Street, N. W., Washington, D.C. 20036.

g. An enlarged copy of "Science Fair Steps," see p. 16, to help students look ahead to the local, state, national and international fairs and encourage them to plan their time and expand their vision.

h. Examples of webbing diagrams (provided) to show students how webbing can be done; they'll have a chance to try it after Brainstorm Bridge.

SUGGESTIONS FOR CLASSROOM MANAGEMENT

To facilitate management of your classroom, offer a variety of diverse delivery systems to accommodate different modes of learning; give equal opportunities for independent study, small group study and entire-class interactions. Set up a learning center devoted to the **SEARCH** unit, providing task cards, contracts, games, teacher-made kits (see above). Arrange the classroom to encourage freedom of movement, cooperation and involvement (see suggested layout below). You'll need space for resources and a spot for the conferences you'll have with each student. A time-out corner is also a good idea– the brain needs relaxation time in order to generate new ideas and problem-solve. For more information see **Optimizing Learning** by Barbara Clark.

Offer your students a way to facilitate their information-gathering tasks: designate a color for each problem or question under their topic; each time they find an answer in their research, they can color-code their note card. Using notecards and different color pens to record this information will help students organize their data easily.

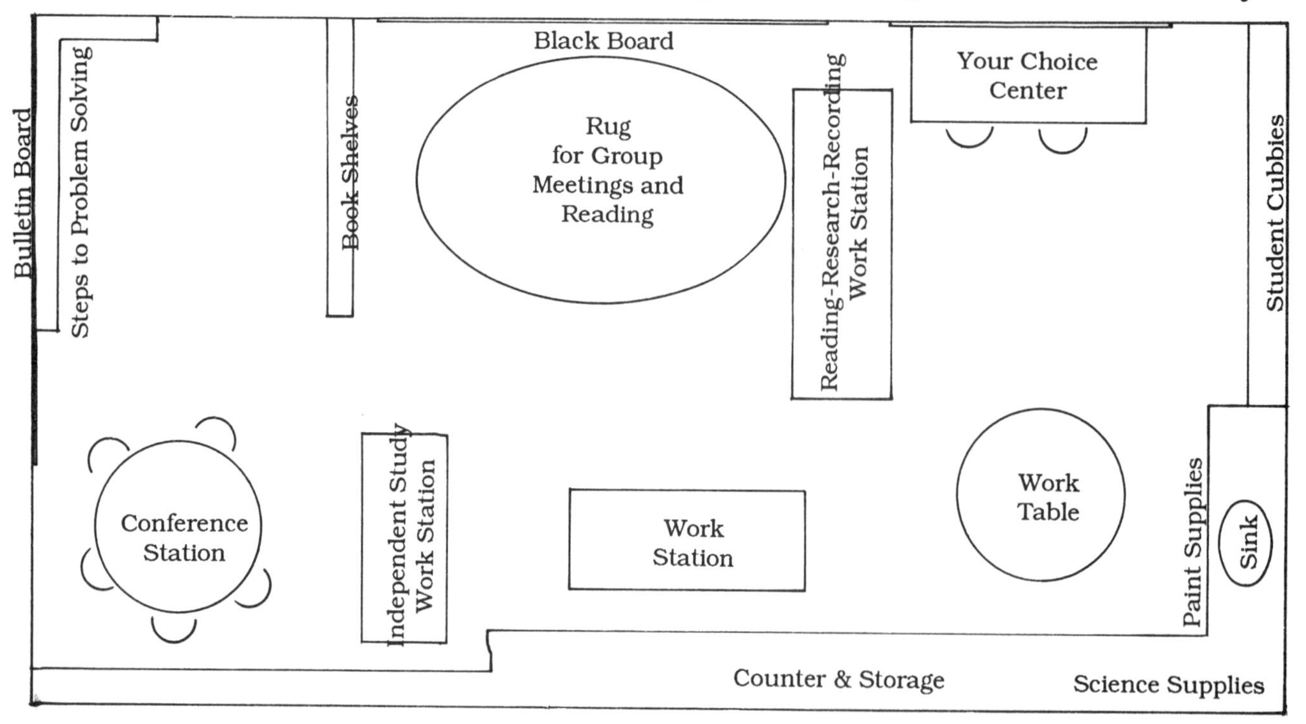

Send for information early in the school year for the Science Talent Search Contest, held each December for high school seniors. Applications and details available from Science Service, 1719 N. Street, N.W., Washington, D.C. 20036. They will send you student application packets. Scholarships are awarded by the Westinghouse Science Awards to contestant winners. See the bibliography for information about past Westinghouse winners (a good resource for your Learning Center).

A field trip to a Public Library or a University Library could be planned to familiarize students with resources not found within school libraries. A trip to a Science Fair provides students with an opportunity to see how projects are displayed, viewed and evaluated. Use the "Letter to Parents" at the end of this section to encourage parents' support, help, suggestions, resources.

Using a copy machine, duplicate the Rest Area Tokens on page 24 and use them as awards to mark your students' progress on the **SEARCH** roadmap. Use colorful, heavy-weight papers (such as Astrobright, available at large office supply stores) to brighten up the students' maps. You could also copy the master map on colored paper and laminate it for display in the Learning Center.

Science research projects and independent study projects can also be offered as an extracurricular or community services activity for afterschool enrichment programs.

Instead of using live animals, it is suggested that you use invertebrates such as crickets, meal worms, or pillbugs. Be wary of using toxic substances. Screen any possibly dangerous materials with a supervising adult.

Here is a list of materials, items and equipment needed to complete the **SEARCH** unit. A list sent home to parents may encourage donations of recyclable or discarded materials:

Nylon thread
Masking tape
Felt-tip marking pens
Coffee cans and lids
Common straight pins
Goggles for safety
Simple tools: hammer, screwdriver, pliers, wire cutters
Jars, pint
Kitchen scales
Batteries
Metric ruler
Milk cartons, 1 quart
Plastic jugs and bottles
Paper clips
Paper fasteners
Paper towels

Single-edge razor blades or X-acto knife
Rubber bands
Sandpaper
Scissors
Aluminum foil
Baby food jars
Cardboard
Cardboard boxes
Cloth, wool, cotton, burlap
Nails
Paints, non-toxic
Paintbrushes
Sodium bicarbonate
Shells
Sugar cubes
Steel wool

Tissue paper
Twist ties
Wire coat hangers
Funnel
Beaker
Wire
Siphon
Soft drink cans
Construction paper
Newsprint paper
Crayons
Compass
Lettering stencils (1", 2", 3", 4")
Posterboard (white & light colors)
3x5" cards

Science equipment as needed by students: balance scales, weights, pulleys, test tubes, beakers, pH or litmus paper, eye droppers, filter paper, rubber tubing, magnets.

Simple chemicals: household bleach, baking soda, detergent, salt, vinegar.

Your Steps to the SCIENCE FAIR!!

International Fair date______ sponsor____________
location__________________

National Fair date______ sponsor____________
location__________________

State Fair date______ sponsor____________
location__________________

Regional Fair date______ sponsor____________
location__________________

Local or City Fair date______ sponsor____________
location__________________
May be entered without winning school fair

School Fair date______ sponsor____________ location____________

PLANNING THE PROJECT

When students reach Brainstorm Bridge on the **SEARCH** Roadmap, some good questions for you to ask are:

What is your purpose?
Who is your audience?
What method of research will you use?
How will you organize your information?
How will you present your findings?

It is important for students to know that conducting an experiment is not the only way to complete this unit; encourage an understanding of the different **methods of research** available to your students. Develop a classroom discussion that will produce ideas for different methods detailed below. Use some of the examples of real projects on pages 28 and 44. When students reach Brainstorm Bridge, refer to the webbing diagrams provided; they were created by the same students who did the ESP study cited on page 28, example 1.

Historical method: a way to do research that describes what has happened in the past. It looks at how, when, and why past events occurred. Read about observations made by people in the past who wrote letters, diaries, journals recording events. Look in archives, history books, encyclopedias, old journals and magazines.

Descriptive method: information is collected by looking at the way things are. Examples of this method are popular opinion surveys (such as the Harris Poll), television and radio rating studies, and political polls. The data collection devices are usually checklists, questionnaires, and interview guides.

Experimental method: there are two ways to set up experimental research. In one, two or more variables are manipulated against a control to determine a cause and effect relationship. In the other, a theory is used to make a prediction (hypothesis), then it is tested.

Correlational or predictive method: statistics or numbers of things (activities, things that happen, incidents) are compared and conclusions drawn. Can be used to establish a trend and then predict future possibilities. This method will utilize graphs, charts and data bases.

Students need to understand how each method works before they choose the one best suited to their hypothesis. After discussing possible methods of research, have students brainstorm possible **products:** an extensive list is included in the student section. Write the brainstormed responses on the chalkboard. Then go over each one and consider what **skills** are needed in order to produce a quality product. The following are listed for your quick reference.

For Media-Related Presentations:

Skills:

Recording on cassettes
Recording on video
Scripting
Storyboarding
Photography
Using sound effects
Selecting background music
Using computers

Mixed Media Skills:

Art skills
Lettering techniques
Layout & design
Using a laminator
Using an X-acto knife
Design construction skills
Dancing skills
Understanding of weights and balances (for building mobiles)

For Written Presentations:

Graphing
Compiling statistics
Layout and design
Using word-processor
Book binding
Making overhead transparencies

For Oral Presentations:

Personal image awareness
Elocution
Use of gestures
Question-asking
Interviewing skills
Using a microphone
Imaging and rehearsing interviews

The following worksheet provides a practice activity for students before they write their final plan. For more on Bloom's Taxonomy and more verbs, see appendix.

PRACTICE WORKSHEET

Name ______________________ **Your Topic** ______________________

PROCESS		**PRODUCT**	
Thinking Level	**Verbs**		**Sample Activity**
Recall and Comprehension: Giving descriptions, stating main ideas (who, what, when, where)	Show Demonstrate Name Illustrate Label Paraphrase Locate List Explain Observe	Chart Filmstrip Map Diagram News article Model Book Illustration	
Application: Solving problems, classifying (a single correct answer)	Construct Organize Solve Experiment Collect Report Demonstrate Sketch	Scrapbook Map Mobile Collage Cartoon Song	
Analysis: Finding evidence, giving opinions, identify-ing motives and causes	Dissect Analyze Categorize Separate Classify Examine Survey Suppose Attribute-listing Compare and contrast	Graph Chart Questionnaire Dictionary Puzzle Survey	
Synthesis: New solutions Original thought Predicting Unique approaches	Infer Imagine Combine Design Formulate Solve Predict Hypothesize Invent Compose Construct	Code Invention New Language Contest Poem Interview Solution	
Evaluation: Judging the worth or validity and supporting the results	Interpret Rank Justify Grade Evaluate Assess Judge Critique	Debate Editorial Chart Display TV Broadcast Letter Demonstration Hall of Fame	

EVALUATION STRATEGIES

Students can be evaluated in the following ways:

1. Final projects are evaluated according to the criteria in the performance objectives, i.e., the report must have all parts organized and concisely written. The project components such as the written report, display, model, or product must reflect the student's in-depth understanding of the research topic, address the hypothesis, and employ one of the research methods such as experimental, non-experimental, or survey. Other considerations for the project are creativity, originality and neatness.

2. The student's completed copy of the book, **SEARCH,** can be another means of evaluation. Students and teachers may find the Independent Study Checkpoints and Independent Study Record (p. 50-51) not only valuable record-keeping tools, but also two methods of evaluating the process of learning for the student.

3. Individual conferences with students while they are working on the steps in **SEARCH** will provide feedback on the student's efforts, quality of work, and their mastery of research skills.

4. A Self-Evaluation form (p.79) should be completed by the student.

5. Since students work to enter a Science Fair or present their project to another appropriate audience, judging and evaluating provide yet another way for the student's work to be assessed. In addition to displaying the project, a personal interview with the student is often involved. Sometimes winners receive ribbons and other prizes for their outstanding work. If the project is presented to an appropriate audience, the student and the audience or certifiers need to know the criteria for evaluating the project.

6. The project must reflect the following considerations:
    a. the student adequately researched the topic, using the appropriate data gathering techniques, and accuracy.
    b. the student presented the hypothesis clearly and the project satisfied the hypothesis.
    c. the student utilized the scientific method and the problem was based on scientific principles.
    d. the student used good record-keeping of experiments and data.
    e. the student's data/results are analyzed and interpreted clearly and logically.
    f. the student compares her findings to other research findings.

g. the student shows creativity in the project by approaching the basic problem in original or unique ways.

When looking at a student's work, it is important to examine what thinking processes were used and applied during the search. Here are some examples of how to question and evaluate the student's thinking. A teacher could have a class discussion about these questions to prompt the students to think about their thinking.

WILLIAMS' CREATIVE THINKING PROCESSES

Did the student select activities that developed:

Fluency: Did the students list as many ideas as they could think of on their topic? Did they brainstorm by themselves, with a partner, with a parent? Did they brainstorm and list ideas? Did they gather all the facts to define a problem? *Fact Finding*

Flexibility: Did the students categorize ideas on their topic? Use webbing? Cluster their ideas? Categorize their data? Categorize the results they logged? Categorize when designing a survey or interpreting the results? Analyze sub-problems and categorize problem parts? Did they ask this type of question: "In what ways might I . . . ?" *Problem Finding*

Originality: In what ways did the students show unique or unusual ways to approach their project? Do students have an idea in their project that no one else has thought of?

Elaboration: Did students use the elaboration matrix (who, what, when, where, how) when they were notetaking? Did students add lots of detail to their research, observations, or product?

SCAMPER SKILLS: A WAY TO THINK CREATIVELY

As the students were thinking up ideas did they:

Substitute Ideas, e.g., substitute a heat lamp for the sun when doing an experiment with solar model home?

Combine Ideas, e.g., align solar collectors to the direction of the sun?

Adapt Ideas, e.g., use a styrofoam cooler to insulate the walls of a solar model home?

Modify Ideas, e.g., modify clear plastic tubing by painting it black to absorb the heat of the sun for the model-home active-solar water system?

Minify, Magnify Ideas, e.g., make the tubing smaller or larger depending on the size of the model solar home?

Put to Other Uses, e.g., find out how many different uses there are for solar energy (drying clothes, heating water, growing plants, heating homes)?

Eliminate Ideas, e.g., eliminate the heat buildup in the solar home by using a fan to circulate the air?

Rearrange Ideas, e.g., rearrange the placement of the solar collectors on the roof of the solar model according to the angle of the sun's rays?

Reverse Ideas, e.g., line the inside of an umbrella with aluminum foil to direct the sun's rays to a center focal point rather than using the outside of the umbrella to shield rain?

Did the student generate ideas that led to a real solution? *Idea Finding*

BLOOM'S TAXONOMY: MORE CREATIVE AND CRITICAL THINKING

Did students develop their topic by using:

Knowledge: can the students define their topic? A student defined and explained passive and active solar energy.

Comprehension: what did the students observe in their experiment or study? A student studying solar energy observed and charted the temperatures both inside and outside of the solar model home as well as the time and weather conditions.

Application: who did the students interview and how did they apply what they learned from the interview to their project? A student interviewed a solar home owner and took a tour of his home; the student then used the idea of circulating air in the model home as he had seen it work in the real home.

Analysis: did the student compare or contrast ideas, data or variables? A student compared sunny day temperatures to cloudy day temperatures to conclude that even on cloudy days there was a lot of solar heat in the solar model home.

Synthesis: what did the student predict? From careful observation a student predicted that even though it might be 20° outside, the solar model home would obtain at least a 60° temperature inside from solar buildup, if it was sunny.

Evaluation: how does the student know that what they did was valid? What criteria were used? Did they solve the problem using the criteria to determine the best solutions? How did they do this? *Solution Finding* The student who built the solar model home made careful observations of temperature, time, and weather variables over 60 days. He took temperatures with calibrated thermometers and logged precise results. The student then took his data base results and interpreted them using the supporting evidence he had found in past research articles. He related his findings to those in other similar experiments.

Did the students develop a plan or decide how to implement the best solution? *Acceptance Finding* What were their suggestions for future experiments or future plans based on their own findings?

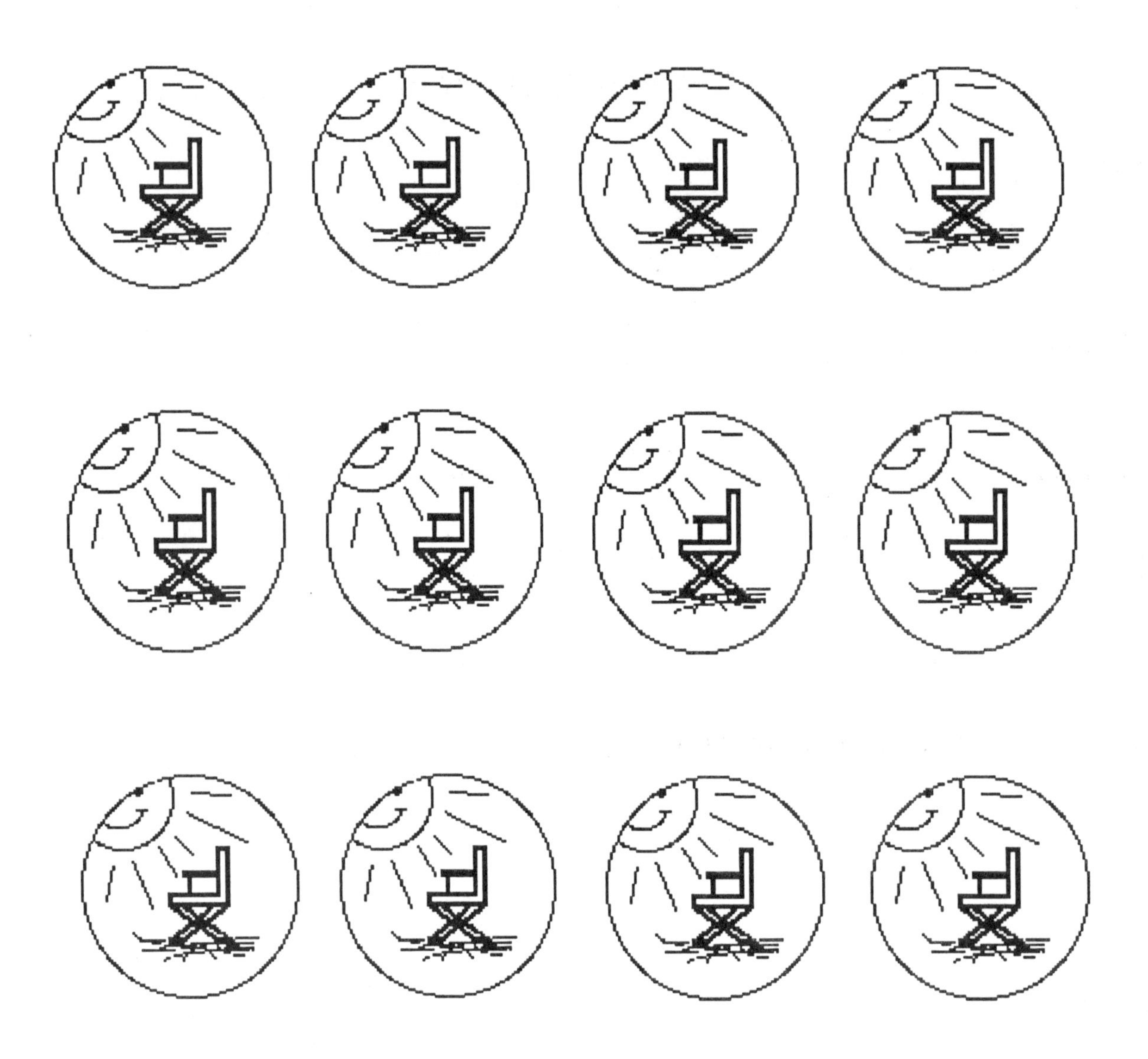

REST AREA TOKENS

Give these Rest Area Tokens to the students after they complete certain steps in the project. A check of the students' work and a conference are best scheduled at each rest area.

Teacher Evaluation

Name: ______________________________

Topic: ______________________________

Comments on Project: ______________________________

Criteria:	Rating lowest				highest
1. Organization	1	2	3	4	5
2. Presentation	1	2	3	4	5
3. Meaningful and understandable	1	2	3	4	5
4. Showed in-depth knowledge	1	2	3	4	5
5. Carried out plan	1	2	3	4	5
6. Effort	1	2	3	4	5
7. Creativity & problem solving ability	1	2	3	4	5
8. Overall rating	1	2	3	4	5

Suggestions for improvement:

____

Things I liked best about this project: ______________________________

SEARCH Lingo

Have students define these words and write the definition next to the word.

1. Hypothesis
2. Webbing
3. Survey
4. Brainstorm
5. Search
6. Notetaking
7. Interview
8. Microfiche
9. Reservoir
10. Display
11. Abstract
12. Periodical
13. Questionnaire
14. Data
15. Solution
16. Evaluation
17. Experimental group
18. Control group
19. Variable factor
20. Technology

Students may need to use a dictionary or glossary.
Now have students take their words and definitions and play **SEARCH LINGO:**

The game is played by students dividing into two teams. Team A reads a Lingo card, which has the definition or description of one of the terms on the **SEARCH** map. Example: "A possible solution to explain your project data; that which is proven by scientific method." The correct answer is "hypothesis."

Team B, the opposite team, must tell what the term is whose definition is being read. Answers can be in students' own words and age-appropriate. If that team answers correctly, then the team is awarded a point for the correct answer. If the answer is incorrect, no point is awarded and the next definition card is read by Team B to Team A, who now has the opportunity to answer for a point.

LETTER TO PARENTS

Today your child ______________________ begins a unit called **SEARCH,** a guided independent study emphasizing the acquisition of research and problem-solving skills. The student will be encouraged to work independently but may need a little extra encouragement and support during the next 8-10 weeks. Here are a few of the ways we can use your help:

- Provide needed transportation to libraries or other sites
- Assist with materials or tools student may need
- Chaperone class field trips
- Offer guidance when students request it
- Volunteer in classroom
- Suggest speakers, refer business or professional contacts or offer your own expertise as an authority

Student's chosen topic is ____________________________________

____

teacher signature

- -

Yes! I will be glad to cooperate with your classroom during the **SEARCH** unit. I will do the following: ____

____

signed, ____

parent

Here are some examples of independent study projects to give you some ideas of the range of possibilities open to students in their **SEARCH:**

1. Two 6th-grade students chose the field of parapsychology for their science project. Their hypothesis was "ESP is 98% tricks or fantasy; 2% real or scientific." They did a survey with 80 students in their school on ESP and computed the results. The two students did 40 experiments on ESP and charted their results. Their conclusions were presented to their teachers and Science classes to be further presented at a Science Fair. (See their ESP webbing charts on p.30.) *Correlational study*

2. One concerned 5th-grade student built a workable burglar alarm for his home after he researched crime in his area. He worked with the local police department to get facts on burglaries. He determined that alarms were a deterrent to home burglaries. He worked with his father to gather materials to build his alarm. He also tested his alarm for reliability. *Experimental research*

3. One high school senior chose to study the psychological factors surrounding retired citizens age 60 and over. She created a survey that she gave to over 100 community members. From that survey she tabulated the results and published the data in the town's newspaper. Her purpose was to make the public aware of the psychological factors of aging as well as how to positively cope with these factors. *Descriptive research*

4. A 5th-grade student designed an experiment to test the effects of aspartame. She received an award for outstanding work from the Dietetic Association and the Osteopathic Physicians and Surgeons Auxiliary for her experimental work. *Experimental research*

5. One 8th-grade girl did an independent study on archaeological digs in the U.S. and how one goes about participating in them. *Descriptive research*

6. A 7th-grade girl did an independent study on jogging. She wanted to increase her running time and in so doing measured the daily effects on her heart rate. She did this study over a period of 6 weeks, keeping a daily log of her running time and pulse rates before and after running. *Experimental research*

7. One 7th-grade student researched *Pi* (the number 3.1416). He had read that the number Pi might repeat itself when it reached the 200th decimal, but it did not repeat itself. He also did historical research on Pi, generating a timeline of when it was first discovered and its history and use to the present. *Historical and experimental research*

8. A high school sophomore who had an interest in computer programming researched and taught himself how to program a voice synthesizer for an elementary handicapped student to use in speech therapy. *Some projects may lend themselves to investigation and questioning. The results would be explored and used to solve a real problem.*

Here is a list of several other winning projects, all done by 4th and 5th graders:

a. How Do We Stop Acid Rain?
b. How Many Items in Your Home Are Hazardous?
c. Which Colors Inspire and Affect Our Moods and Feelings?
d. How Do Dogs Communicate?
e. Does Vitamin C Prevent Colds?
f. What Prevents Heart Disease?
g. How Much TV is Good or Bad?
h. The Migration Patterns of the Killer Whales

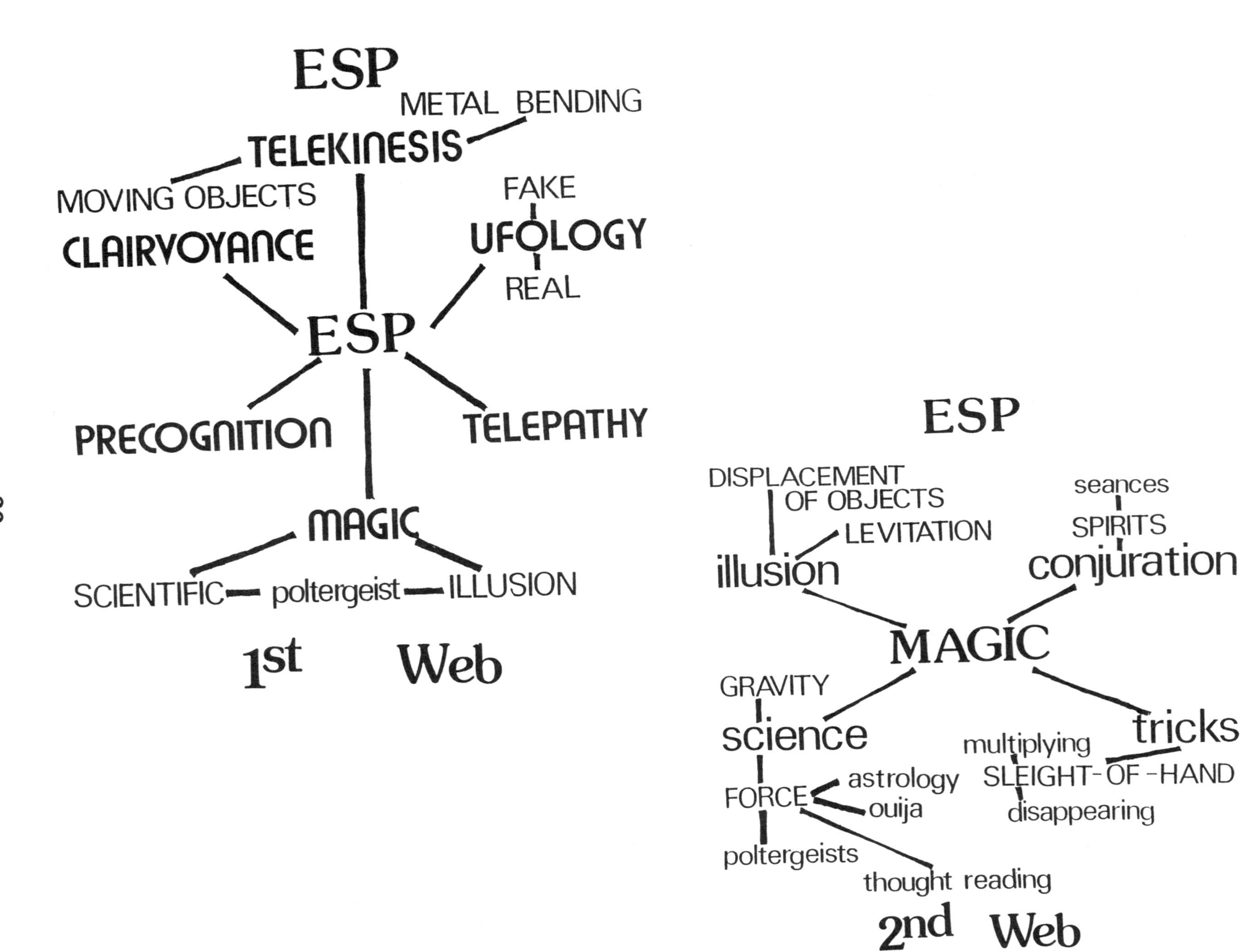
ESP
METAL BENDING
TELEKINESIS
MOVING OBJECTS
FAKE
UFOLOGY
REAL
CLAIRVOYANCE
ESP
PRECOGNITION
TELEPATHY
MAGIC
SCIENTIFIC
poltergeist
ILLUSION
1st Web
ESP
DISPLACEMENT OF OBJECTS
LEVITATION
illusion
seances
SPIRITS
conjuration
MAGIC
GRAVITY
science
tricks
multiplying
SLEIGHT-OF-HAND
disappearing
FORCE
astrology
ouija
poltergeists
thought reading
2nd Web

PART II
FOR THE STUDENT

TABLE OF CONTENTS FOR PART II FOR THE STUDENT

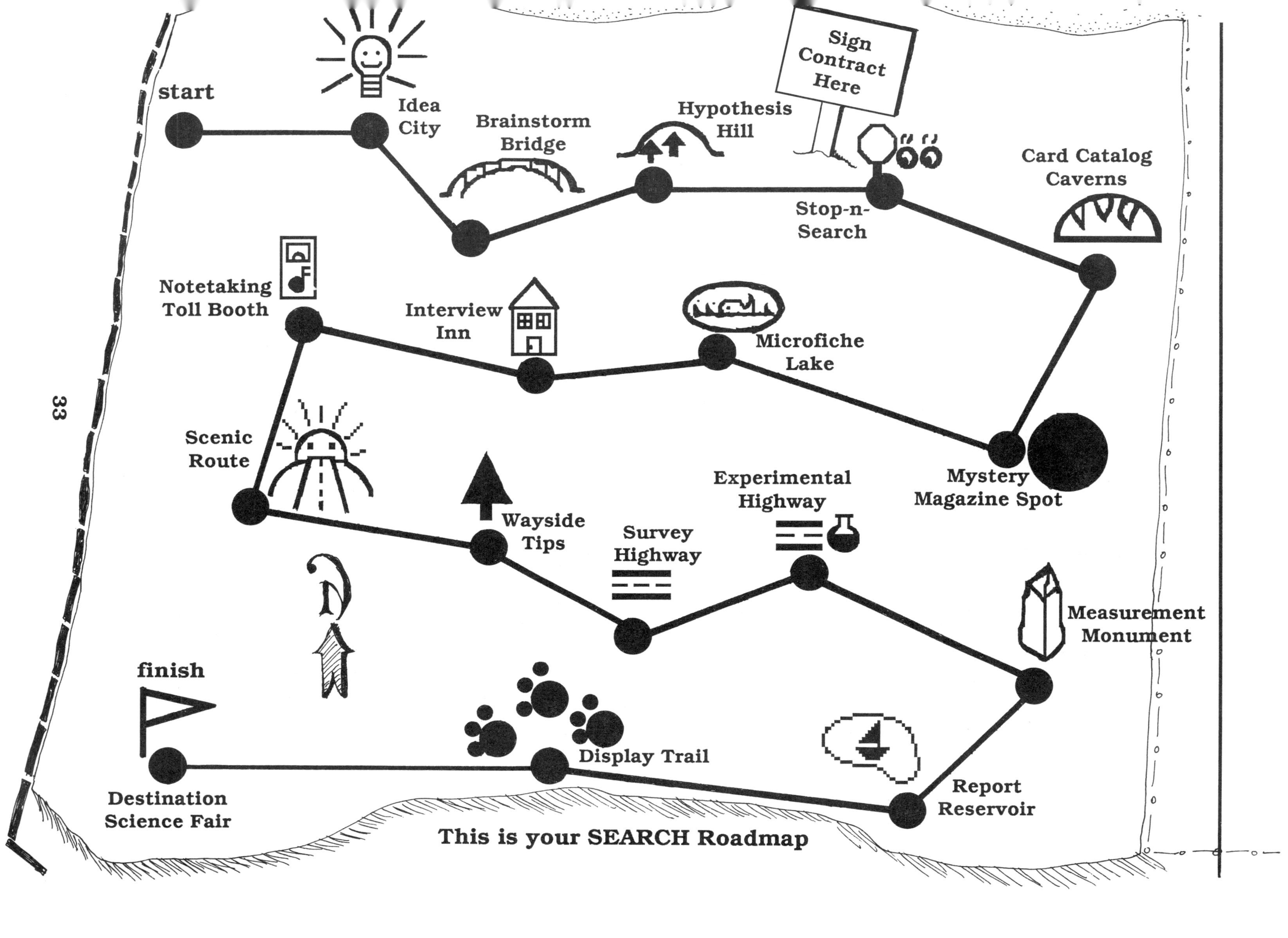
start
Idea City
Brainstorm Bridge
Hypothesis Hill
Sign Contract Here
Stop-n-Search
Card Catalog Caverns
Notetaking Toll Booth
Interview Inn
Microfiche Lake
Mystery Magazine Spot
Scenic Route
Wayside Tips
Survey Highway
Experimental Highway
Measurement Monument
Report Reservoir
Display Trail
finish
Destination Science Fair
This is your SEARCH Roadmap

INTRODUCTION

SEARCH is a guided tour of the research processes necessary to complete an Independent Study or Science Fair Project. The purpose of an Independent Study is to create a project that has a purpose and is useful. It may also help you to explore a science career, study a topic in-depth, or discover new areas of interest and concern.

On the previous page is the **SEARCH** Roadmap, specially designed to take you, step-by-step, through all the things you need to do to complete a successful project. At each stop along the way, there are new words to learn, assignments to complete, and worksheets for practicing what you've learned. Any time you need help, be sure to refer back to the beginning part of your packet, or ask your teacher for a conference. Use the Learning Center available to you in your classroom, and all the other resources offered by your teacher and through your library. Be sure to stay on the road and pause at each rest stop; you are off on an exciting and productive adventure!

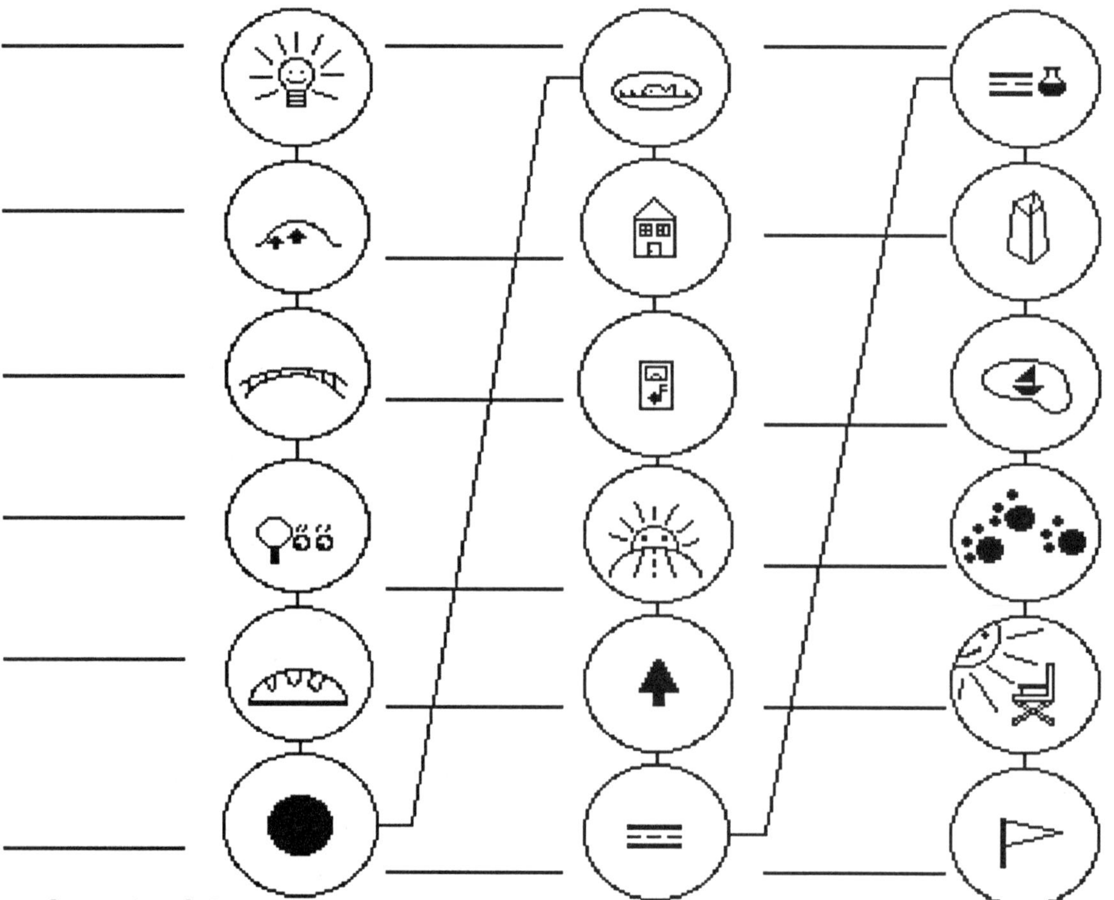

Identify each of the symbols on the **SEARCH** Roadmap. Write the answers on the lines.

Fill up these circles with your tokens that you receive along the way. See your teacher for a token or signature at each site after you finish each step in your search.

REST AREA TOKENS

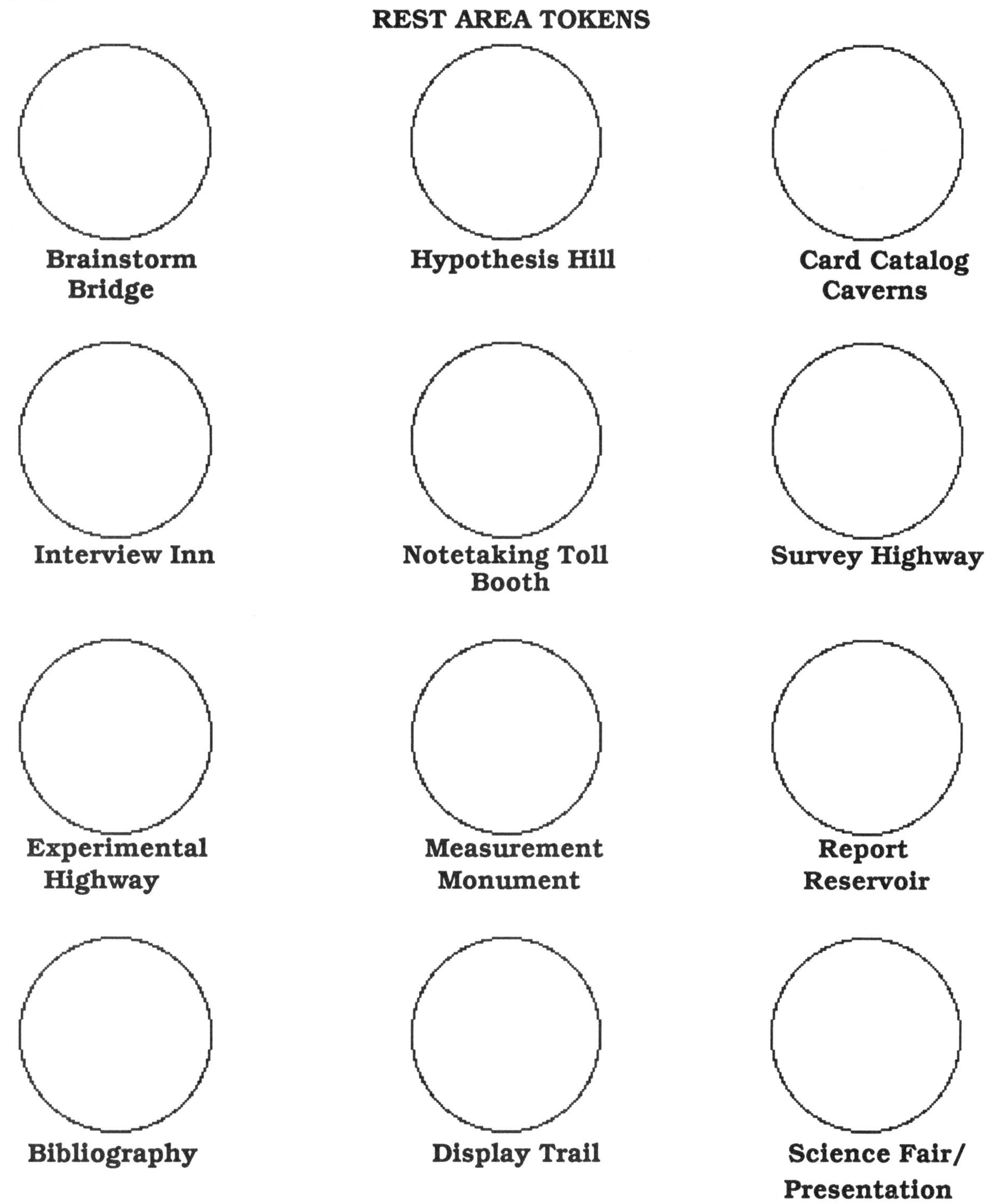

The Journey Begins

A successful research plan begins with three things:

1) a topic that interests you

2) a list of the steps you'll take

3) an idea for a useful product or outcome

To help you begin step 1, stop at Idea City (p. 37), where you will take an interest survey to get the ideas flowing.

To help you begin step 2, complete the Independent Study Contract and the Independent Study Checklist after Hypothesis Hill.

To help you begin step 3, look at the Product List (p.45) to help you decide on a product.

Idea City

You are now beginning **SEARCH**, a journey to learn more about a special interest of yours. The purpose of this questionnaire is to help you think about your special interests. Since your interests are yours alone, your answers here will be different from anyone else's. The right answers are the ones that fit you best.

Read through the questions before you begin. Take time to think about each one; then go back and fill in your answers.

1. When I imagine myself as a famous person, I see myself being ______________

____

____

2. When I visit, or imagine I visit, a museum with every kind of exhibit, I go first to see____

____

____

I don't mind if I miss the exhibits of ________________________________

____

3. If I could fly anywhere in the world, my first choice would be ______________

____

My last choice would be ____ because ____

4. When I think about all the people I've read of or know about from the beginning of time till now— the person I would most enjoy visiting is ____________

____

I would like to talk to that person about ______________________

____

5. My favorite books are about ______________________________

____

6. If I had a collection of items that were special to me, I would collect:

Mark your first choice with a **1**, second choice with a **2**, and third choice with a **3**. Circle any collection that you already have.

_____ Stamps

_____ Telescopes

_____ Programs from plays

_____ Baseball cards

_____ Musical records or tapes

    _____ symphonies

    _____ rock

    _____ jazz

    _____ opera

_____ Fashion magazines

_____ Butterflies

_____ Paintings

_____ Weather vanes

_____ Pottery chards

_____ Historical facts

_____ Seashells

_____ News clippings

_____ Miniatures

_____ Postcards

_____ another idea: ______________________________

7. If I had the ability to do any of the following, I would choose to:

Mark your first choice with a **1**, your second choice with a **2**, and your third choice with a **3.**

Become a member of an archaeological dig _____

Perform in a theatrical event _____

Publish a poem or other writing _____

Create a song _____

Design a building _____

Plant a tree _____

Conduct an orchestra _____

Become an astrologer and study the stars _____

8. When I think about school and what I like to learn about, I'm most interested in

____

9. If I could write a letter to anyone in the world, I would write a letter to _______________________and ask_______________________

____

10. When I think about the world in general I would like to know more about

____

11. If I could learn about anything in the world I would choose the subject of

____

The main question I would like to answer about my subject is _______________

____

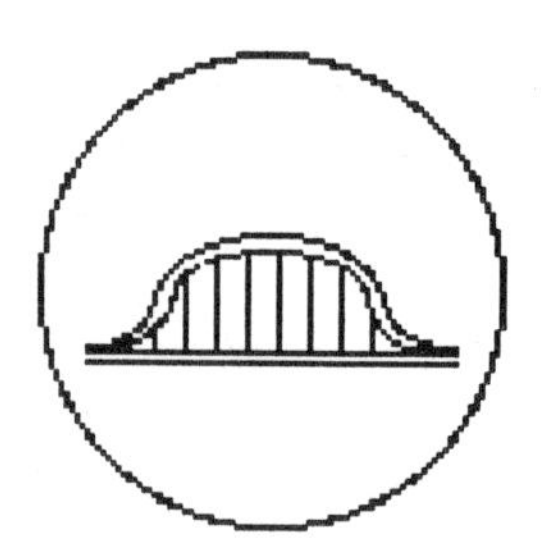

Brainstorm Bridge

Brainstorming is a way of gathering lots of ideas.

Use these simple rules when brainstorming with a partner or a group of friends:

1. Accept every idea.
2. Lots of ideas; the more ideas the better.
3. Zany ideas are great!
4. Piggy back ideas— one idea helps create another one.

Think of at least 15 questions about your topic.

Example: Why is solar energy less expensive for heating buildings than other fuels?

My 15 questions on my topic are:

1.____

2.____

3.____

4.____

5.____

6.____

7.____

8.____

9.____

10._______________________________________

11._______________________________________

12._______________________________________

13._______________________________________

14._______________________________________

15._______________________________________

Rest Area

Webbing

Another way to look at a topic and get more ideas and categories about your topic is to create a web like this:

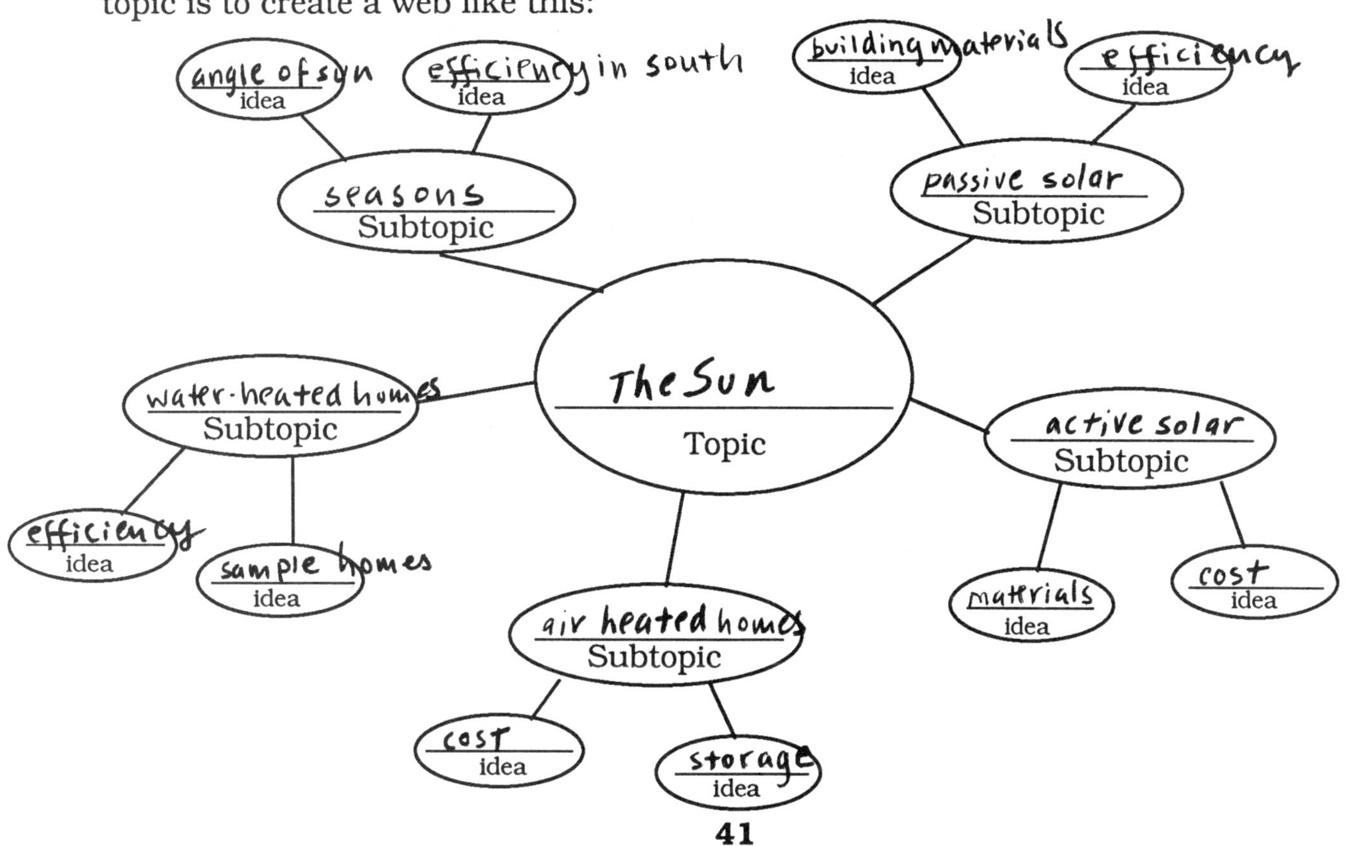

Create another web with your own topic, using the diagram below; start in the center and fill in your topic. Think of subtopics or categories that relate to your topic. Then add ideas to each topic; remember your brainstorming. Add more circles if you wish. When you are finished, you may want to try a second web, building from one idea or subtopic on your first web. Ask your teacher to show you some more examples of webs that students like yourself have made.

You can use your web to make up research questions.

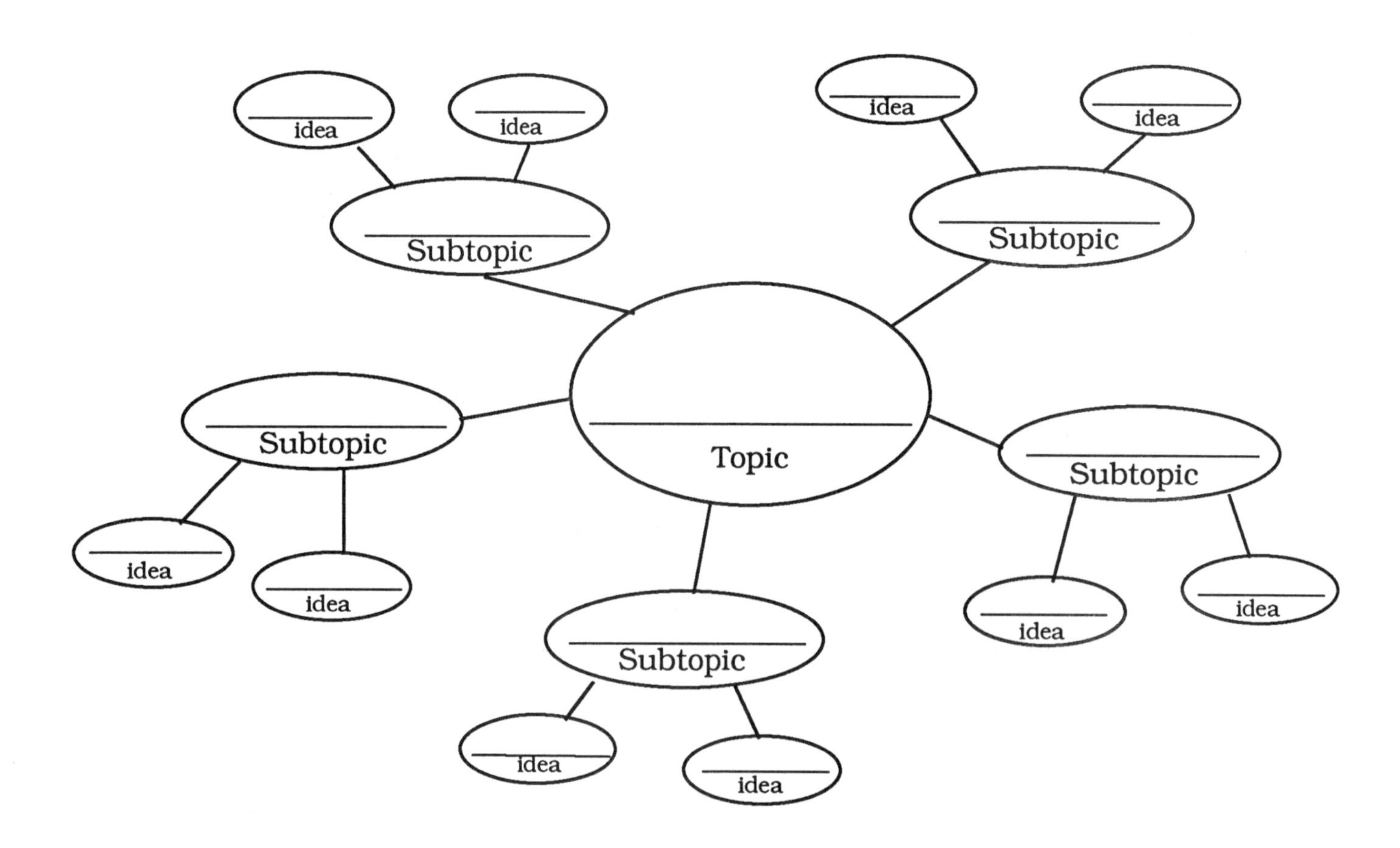

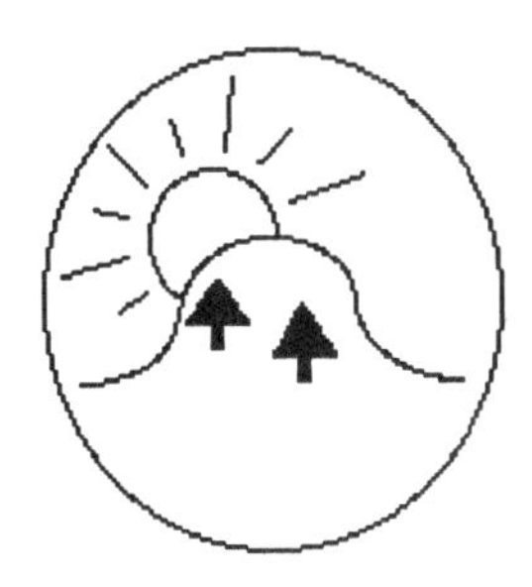

Hypothesis Hill

A hypothesis is a possible solution to a problem or a guess as to what the outcome of the project will be.

Take your question in number 11 from Idea City and turn it into a statement.

Example: Question: Is solar energy less expensive than other fuels?

Statement: Solar energy is less expensive than other fuels.

or

Question: Do colors affect our feelings and emotions?

Statement: Colors affect our feelings and emotions.

Your statement:____

____

____

____

Rest Area

Choosing Your Product

A product is what you'll have at the end of your **SEARCH**. It is putting your ideas, methods, and research into a form everyone can see. It can be **useful** or help to **solve a problem.**

Here are some **useful** products:

A **debate** on new evidence that has been discovered. For example, the migration pattern of killer whales is not really a consistent pattern; it is a random pattern.

A **collection** (such as shells or rocks) that is accurately labeled with researched findings from the area in which they were found. This could be displayed at a local museum, nature center, or mall.

A radio or TV **commentary** about a science issue that has been researched, such as wildlife management, life support systems, substance abuse or organ transplants.

A **presentation** of a Science Project or Independent Study to an interest group relating to the subject, such as an Astronomy Club, Hot Air Balloon Club, Archeological Society, or Audubon Society.

A **slide show** showing solar collectors currently installed in your community.

Here are some products that help **solve a problem:**

An **appearance** or a **letter** read at a legislative hearing concerning a piece of legislation that the student has researched and either opposes or supports.

A **letter** to the newspaper editor on how toxic waste is affecting an area; backed by research and providing solutions to the problem.

A product can be as simple as a written report or as complex as a videotaped documentary tracking your project from start to finish. Look at the list on the following page. Choose three products that you have made in the past, then, choose three products you might like to try to make. Using your statement from Hypothesis Hill, choose three products from the following page that would best carry out your idea. Use a separate piece of paper for this exercise.

Product List

Advertisement
Article for a science journal
Art show
Bibliography
Board or card game
Book
Chart or graph
Collage
Collection
Comic strip
Computer program
Crossword puzzle
Dance
Data base printout
Demonstration
Display
Diorama
Dictionary or glossary, illustrated
Diary or journal
Experiment and/or demonstration of experiments
Film or filmstrip
Futuristics model
Hidden picture
Illustrated story
Information diagram with key
Invention
Invitation to a guest speaker
Guest speaker
Greeting card
Jigsaw puzzle
Labelled diagram with questions
Letter to the editor
Light show for planetarium
Limerick or riddle
Magazine or newspaper
Map with legend
Mini-center
Mobile
Model
Mural
Museum
Musical composition or instrument
Oral report or speech
Overhead transparency
Pamphlet or brochure
Panel discussion
Paper folding
Pattern with instructions
Photo essay
Photo album
Play, dramatization or skit
Poem
Portfolio of drawings or other artwork
Poster or bumper sticker
Puppet show
Radio show
Recipe
Rubbing
Sample specimens
Scavenger hunt
Scrapbook
Sculpture (found objects)
Seek and Find game
Slide show
Song (original) or collection of songs
Tape recording
Terrarium
Timeline
Travelogue
TV program or documentary
Written paper
Video

Now that you've settled on a topic and cleared it with your teacher, fill out and sign this Independent Study Contract:

Independent Study Contract

Name: ______________________________Grade:____________

Today's Date ___________ Planned Completion Date ___________

Project Title ____

Statement of the problem to be researched:

____

____

Describe your project (include goals and objectives you hope to accomplish):

____

____

____

____

My product will be:____

Steps I will follow:

1. Resources I will use (books, magazines, people, etc.):

____

Technical assistance from:

My audience will be:

Special Equipment and materials I may need:

Signed ______________________ ______________________
student teacher

(copy to parents) ______________________date

You are now ready to continue the **SEARCH** roadmap, your guide to learning how to create, design, research and produce a successful science project or independent study project. Remember to review the introductory material in your packet as needed, and consult with your teacher, advisor or mentor. You will have some more practice as you fill out the Independent Study Work Sheet on the next page.

Independent Study Worksheet

Name ____

1. Topic ______________________________________

Subject Area: (choose one or more)

Language________ Art ________ Social Studies__________

Music __________ Science ____________ Math ____________

Other _________________

2. Resources (people, places, books, filmstrips, art, music, magazines, etc.)

what or who where

________________________ ____________________________

________________________ ____________________________

________________________ ____________________________

________________________ ____________________________

3. The method of research______________________________

4. The product____

____

5. Materials and technical assistance needed

what or who where

________________________ ____________________________

________________________ ____________________________

________________________ ____________________________

________________________ ____________________________

6. Timeline

date

date

date

7. Possible audiences

who | how will I contact them?

8. How will my project be evaluated?

9. Optional: An imaginary drawing of what my project will look like.

Use this worksheet to record your daily progress; use the checklist on the following page for the tasks you'll need to write on this sheet. This will also serve as a way for you and your teacher to evaluate your project at the end of **SEARCH.**

Independent Study Daily Record

Name____

Today's Date _______________ Working on Step _____ Completed Step ________

- -

Today's Date _______________ Working on Step _____ Completed Step ________

- -

Today's Date _______________ Working on Step _____ Completed Step ________

- -

Today's Date _______________ Working on Step _____ Completed Step ________

- -

Ask your teacher for more blank worksheets.

Independent Study Checkpoints

1. Brainstorming/Webbing topic________

2. Hypothesis________

3. Contract on p. 46 filled out__________

5. Library research: periodicals, microfiche, card catalog, vertical files, science books_________ Computer modem search___________

7. Interviewing_____________

6. Notetaking___________

8. Gathering: by Survey________________
by Questionnaire_____________
by Experiments____________

9. Use of charts, graphs and tables to report your results

10. Reporting: by term paper_____________
by oral/speech presentation__________
by product______________

10. Bibliography: at least _____ sources listed

11. Abstract: written paragraph or two about your project__________

12. Final presentation: at Science Fair, to a judge, teacher, class or other audience__________

13. Evaluation of your independent study___________

14. Teacher Evaluation of independent study_________

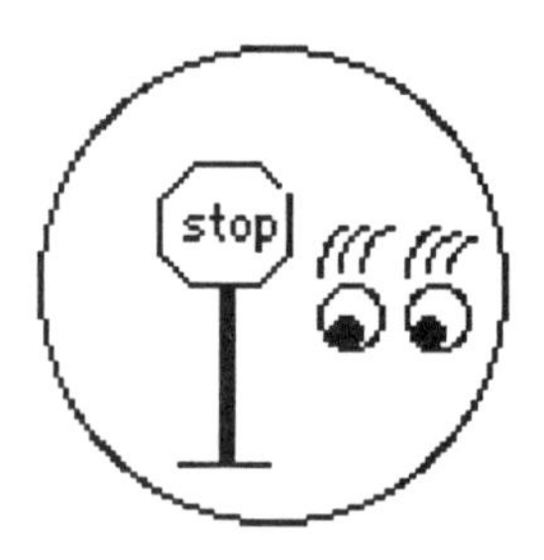

Stop-N-Search Site

Where do I look for information about my topic?

Good question! Just go through the steps that follow. At each point you will need to do something. Be sure to stop at the Rest Areas and get your token or teacher's signature.

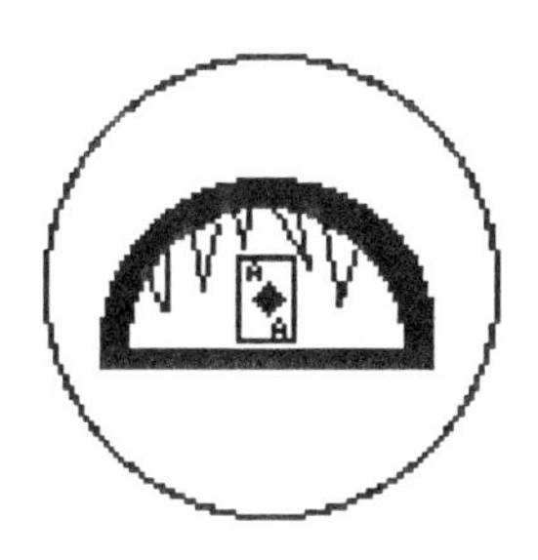

Card Catalog Caverns

At the library, look up your topic in the Card Catalog: usually a big cabinet with lots of little drawers with cards listing all the books in the library. For each book there is a subject card, author card, and title card. List below books you find that may give information about your topic.

Encyclopedias are another place to look; most libraries have many kinds of encyclopedias. Be sure to look at more than one kind.

Here are some other places to look: Dictionaries, Almanacs, *Roget's Thesaurus*, Atlases, *Barlett's Familiar Quotations*, *Physician's Desk Reference*, *Gray's Anatomy of the Human Body*, *Thomas' Register of American Manufacturers* and textbooks concerning the topic you have chosen.

When looking up information in the card catalog, or in any reference books, begin by looking up the subject. Then, choose key words in your subject and look them up. While using any reference materials, remember to use the skills you've already learned, such as your A-B-Cs, how to read a table of contents, index and running head.

Keep a record of ALL your materials on 3 x 5 index cards.

Now, select the books or materials you listed, and begin to read them.

What about a Computer Search? If you can get a search for your topic done on a computer modem that links to a data base or university library, you will have a listing of places to look for your topic. Ask your librarian or computer teacher.

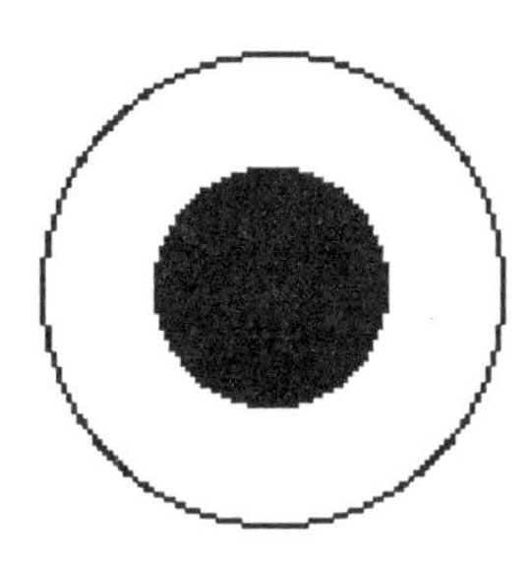

Magazine Mystery Spot

Look in the *Reader's Guide to Periodical Literature,* a book listing all the articles written on every subject that have appeared in magazines and newspapers. There is a separate Reader's Guide for each year.

Look under the name of the subject you are studying.

Remember to keep a 3 x 5 card for each article you read listing the important facts you want to remember. List some of the magazines, newspapers and other periodicals that have articles about your topic.

____

____

____

Don't forget to look up an author who appears to be an expert on your topic.

Other places to look:

Vertical files: contain booklets, pictures and material on your topic.
Anthologies: collections of articles or excerpts from books.
Diaries or journals: document events or personal information.
Catalogs: provide further sources of information, equipment, media.

Microfiche Lake

Ask your teacher or librarian to show you how to use the microfiche films and the reader. Use the microfiche reader to find out if the books, newspapers, or magazines you want are in this library.

You will discover other sources of information on your topic.

List them below:

____

____

____

____

____

Interview Inn

Who are you going to interview?

Think of some people who might know something about your topic (or who might know someone who knows): A scientist, a university professor, a professional in your field, a teacher, an expert on your topic, your parents or other family members, community members, public officials, newspaper editors, corporate managers, an author . . .

Can you think of any others?

Here are some tips:

Call or write to them to set up a convenient time to meet.

Tell them exactly why you want to interview them and what you hope to have them talk about in the interview.

Some possible questions to ask:

How did you first become interested in this subject?
What special training or education did you pursue to learn about your subject?
Do you enjoy working in your field?
What is your opinion on the topic?

Questions that I want to ask are:

____

____

____

Other information that I'd like this person to explain (think about how they might help you show your findings):

____

____

____

Tips for Interview Inn:

- Be a good listener.

- Take notes. Tape recording the interview is helpful. Be sure to ask permission of the person you are interviewing before you use the tape recorder.

- Thank the person you interviewed; a follow-up letter would be thoughtful especially if it includes information about the final aspects of your project (like where it is being displayed, etc.).

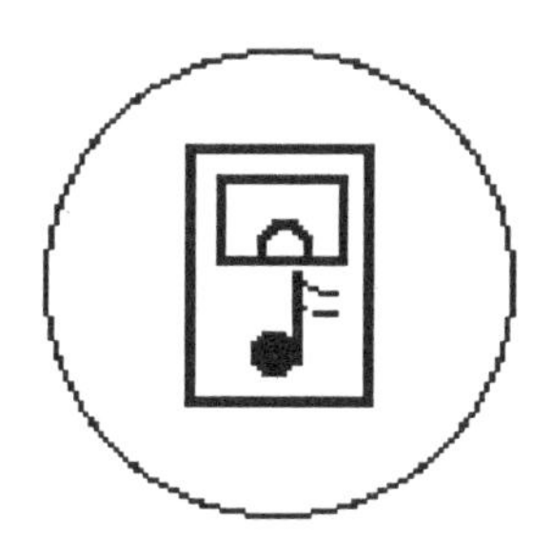

Notetaking Toll Booth

Along the way write out a notecard for each source you use.

Sample:

TITLE:	Brain Power, Secrets of a Winning Team
PUBLISHED:	1984
AUTHOR:	Pat Sharp
PAGES USED:	pp. 20-21
NOTES:	We are born with all the brain cells we'll ever have. The size of the brain changes when we grow. The size of the brain is not related to intelligence

For periodicals, write date, volume and number like this:

6-10-87, Vol. 2 Science News

Stop at the Rest Area when you are finished with all your notetaking.

Rest Area

Notetaking Worksheet

This is a practice sheet to help you search for details as you continue your research. Use it as you read one of your source materials.

Topic of Research:____

Source (reference, book, magazine article):______________________________

Fill in the following:

Who ____

What ____

Where ____

How ____

When ____

Scenic Routes

Here are more places to look for research information. Choose several. Ask a librarian if you need help.

The New York Times: weekly science section (appears on Tuesdays)
Science Index
Scientific American Index
Ayer's Dictionary of Publications (good source of scientific journals)
Ulrich's International Periodical Directory (has foreign science work listed)

ABSTRACTS in science journals are special short reviews of the science articles. This will save you time as you can read about many articles and then choose only the ones you need.

Example: *Wood Energy Resources in Georgia.* Shirley, A. R. 1979. Univ. of Georgia, Institute of Government, Athens, GA 30620.

The current status of wood as a fuel is assessed with a forecast for economic choices giving a larger share of energy fuels to wood in the future. The author foresees combustion processes being developed that will clean up the offensive air pollution products that inhibit wood usage now.

– from *Energy Research Abstracts*

When you finish your project, you will write your own abstract (p. 78). Here is the way an abstract is written:

> Robert Smith
>
> Age 11, Grade 6
>
> Maple Middle School, 15 North Street, Lansing, MI 48917
>
> Science Project: "Is Active Solar Heating Efficient in Michigan?"
>
> For my science project I built a scale model active solar home to determine if active solar energy was an efficient method of heating a home in Michigan. During January, February and March I conducted a test to measure the daily temperature inside and outside the model home. I also recorded the time of day, weather conditions, and the angle of the sun. My tests showed that even on cloudy days there was still solar heat buildup inside the home. And, that the solar collector was more efficient if its face was directly pointed at the sun with no clouds or trees interfering.

Another way to gather information about your topic is to write to the author of a book or article you have read. Here's a sample letter:

(address of person you're writing to)

(date)

Dear _______________,

I am a student at ______________________ (school) and I am doing a research project on ____. I am especially interested in ____.

(the particular focus of your project)

Would you please send me any information you have on this topic that might help me? Please send this material to: ____________________________________

____.

(your name and address)

Sincerely,

(Your name)

When requesting information from an expert in your field, it may help to narrow your question to the particular focus of your project. This way, the expert or author will be better able to provide exactly the information you'll need.

Other places to write for your topic:

1. Science Service, 1719 N. Street, N.W., Washington, D.C. 20036, for the booklet "Thousands of Science Projects"; $3.00.

2. U.S. Supt. of Documents, Government Printing Office, Washington, D.C. 20402

3. National Science Teacher's Association (NSTA) for their Publications List to purchase selected books on your topic. Write to: Special Publications, NSTA, 1742 Connecticut Ave., Washington, D.C. 20009.

4. Look in the *Thomas' Register of American Manufacturers* (a reference book in the library). It lists names, addresses, and phone numbers of all manufacturing companies in the U.S. You can write or phone a company to obtain information on your topic as it pertains to that company.

5. National Aeronautics and Space Administration Jet Propulsion Laboratory, California Institute of Technology, Pasadena, CA 91103 for booklets on physical science topics.

6. National Oceanic and Atmospheric Administration, U.S. Dept. of Commerce, 11420 Rockville Pike, Rockville, MD 20852 for booklets on ecology topics.

7. Public Health Service, U.S. Dept. of Health, Education and Welfare, Rockville, MD 20857 for information on disease and health topics.

8. Nuclear Energy Experiments, Thomas Alva Edison Foundation, 18280 West 10 Mile Rd., Southfield, MI 48075 for free booklet "Nuclear Energy Experiments for Students."

9. American Chemical Manufacturers Association,1825 Connecticut Ave., N.W.; Washington, D.C. 20009 for information on safety.

10. Animal Welfare Institute, P.O. Box 3650, Washington, D.C. 20007 for booklets on animal/biological projects.

11. National Society for Medical Research, 1029 Vermont Ave. N.W., Washington, D.C. 20005 for information on topics of medical research.

Write or phone local community resources such as:

zoos	weather bureaus
museums	hospitals
planetariums	businesses
historical societies	airports
universities	govt. agencies
botanical gardens	newspapers

Ask their library research department for any news articles on your topic.

Look in your local phone book for a listing of local resources.

Wayside Tips

- Look for historical as well as current accomplishments in your field.
- The microfiche is a good source of older newspaper or journal articles.
- Use copying machines to duplicate articles or pages you want to read at a later time or for materials that can't be checked out.
- Be organized. Keep your file cards, notes and information together.

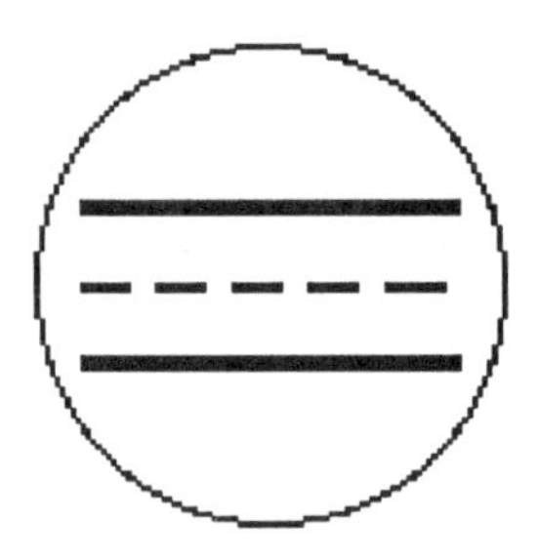

Survey Highway

For your project, you could choose to do a survey. A survey is a way of gathering and measuring many ideas, facts, or opinions.

Here are some survey questions to think about first. Read the three examples before you decide which type of survey you want to do.

Observation Survey: Gathering observations: things we can see and record. Many observations are needed to be accurate. Examples:

1. Observe how you spend your free time every day for a week.
2. Observe how many students eat hot lunch/cold lunch on certain days.
3. Observe number of times one has to mow the lawn in relation to rainfall.
4. Observe the average cost of meals at the local fast food outlet.

Questionnaire: Gathering responses to a question. You can write a questionnaire to find out and distribute it to collect data. Many responses are needed to be accurate. Remember that the more people you survey, the more valid your results will be. You will also have more results to collect and manage. Things to think about:

1. What information do you want to find out by doing the survey?
   Example: I want to know who spends more time studying, boys or girls, and why.
2. Who and how many will be surveyed?
   Example: I will give my questionnaire to all the boys and girls in my class, which is about 50 students.
3. What do you want to ask those surveyed?
   Example: I will ask how long each person studies each day (minutes). I will ask when each person studies each day.

4. How will you do the survey?
   Example: I will give out the questionnaire in person at school; or I will call each person or mail a survey to each person. I will collect the surveys at school during the first hour.
5. How will you record your answers?
   Example: I will use my notepad to record answers. Or I will make a chart that records the answers in columns. I could use a tape recorder if I call the person. Finally, I could write a computer program that tabulates my answers and prints out the results.
6. How will you report the results after you have collected and analyzed them?
   Example: I could use graphs such as bar graphs, line graphs, circle graphs, or pictographs. Or I could use a computer printout that shows the results in this way: the response from each student to each question, then the total results, such as 28 boys responded that they studied 20 to 30 minutes daily.

Review of the Literature: Read surveys that have already been taken and compiled and determine if that data is related to your topic. This can be public opinion surveys or newspaper surveys.

Example: Read about the number of whales that are listed in existence each year in international waters for a period of ten years. Then read about how many whaling businesses have fished in those areas. Draw correlations between the two numbers.

Does the number of whales show a real decline?

Year	Estimated No. of Whales	Whales Caught for Business

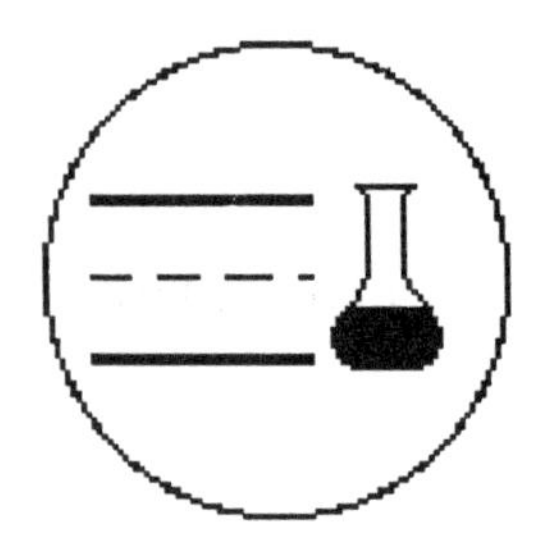

Experimental Highway

Your experiment must answer your question or test or show your findings by:

a. Setting up a control group that will receive nothing different in the testing conditions.

b. Setting up an experimental group that will be the test group.

Sometimes you will need one Control group and many Experimental groups with different factors in each group.

Hypothesis: Plants grow faster when exposed to music.

The constants in the experiment would be type and size of plant, temperature, lighting, food and water. In the experimental group the music would be played at the same volume, at the same time of day, and for the same amount of time.

When this particular experiment was set up, a group of 6 plants were placed in a room with no music. Two plants were in a room with classical music, two with jazz, and two with rock.

Answer the following questions about your experiment:

1. What is one thing you will do to cause a change in your experimental groups that is different from the control group?

2. How many times will you do this to create change?

3. When will you make this change?

Keep a written record of everything you do. Create a log of experiments like this:

OBSERVATIONS AND DATA		
Date	Control Group	Experimental Group

You could also use your computer as a tool for keeping a daily log of experimental data.

"Testing my Hypothesis"
Experiment Memo

Topic of my experiment:______________________________

Background: What ideas or facts do I already know about this topic?

____

Question: What do I want to know? What will my experiment tell me?

My prediction: ____

Planning: The steps I'll take to complete my experiment are:

1. ____ ____________
teacher initials

2. ____ ____________
teacher initials

3. ____ ____________
teacher initials

4. ____ ____________
teacher initials

5. ____ ____________
teacher initials

Results:

What happened in my experiment?

____

____

Reporting: How will I report the data (graph, table, notes)? ____________________

I got these results because:____

____

____

My prediction was the same as or different from the results:

____

____

If I did this experiment again, I would do these things differently because:

____

____

Conclusion:

____

____

____

____

____

____

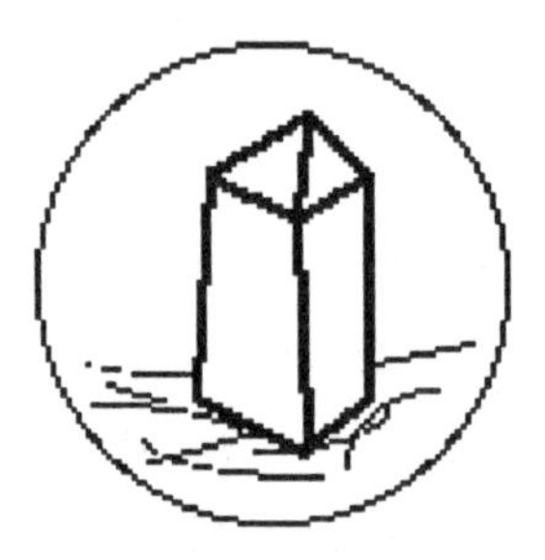

Measurement Monument

One of the best ways to see the results of your experiment, and what they mean, is to put them in graphs. There are lots of graphs: pictograph, bar graph, line graph, and circle graph.

Here are some ways to use graphs:

Bar Graph (can be horizontal or vertical; compares amounts)

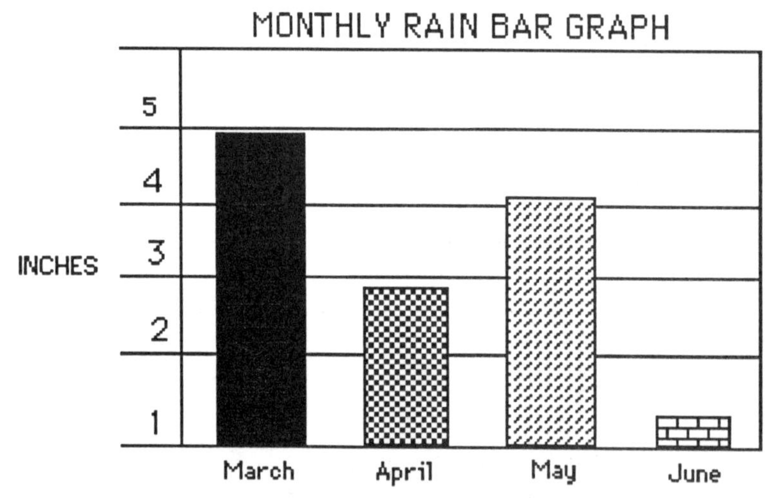

Line graph

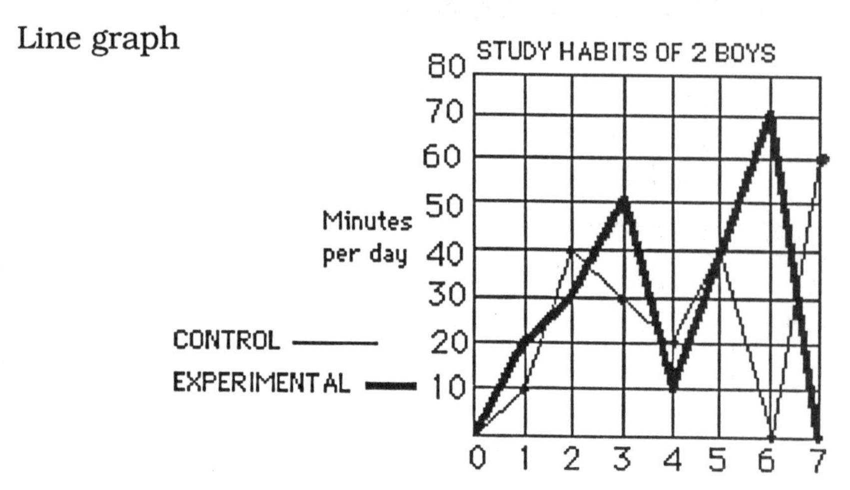

Pictograph

Example: To discover how many baby birds there are in a quarter-mile area around your house during March, April, and May, you could list the names of birds and their pictures. Each time you see a baby bird of that species record it and also where you saw it. Take a walk to discover the nests and locations of birds so you can observe them.

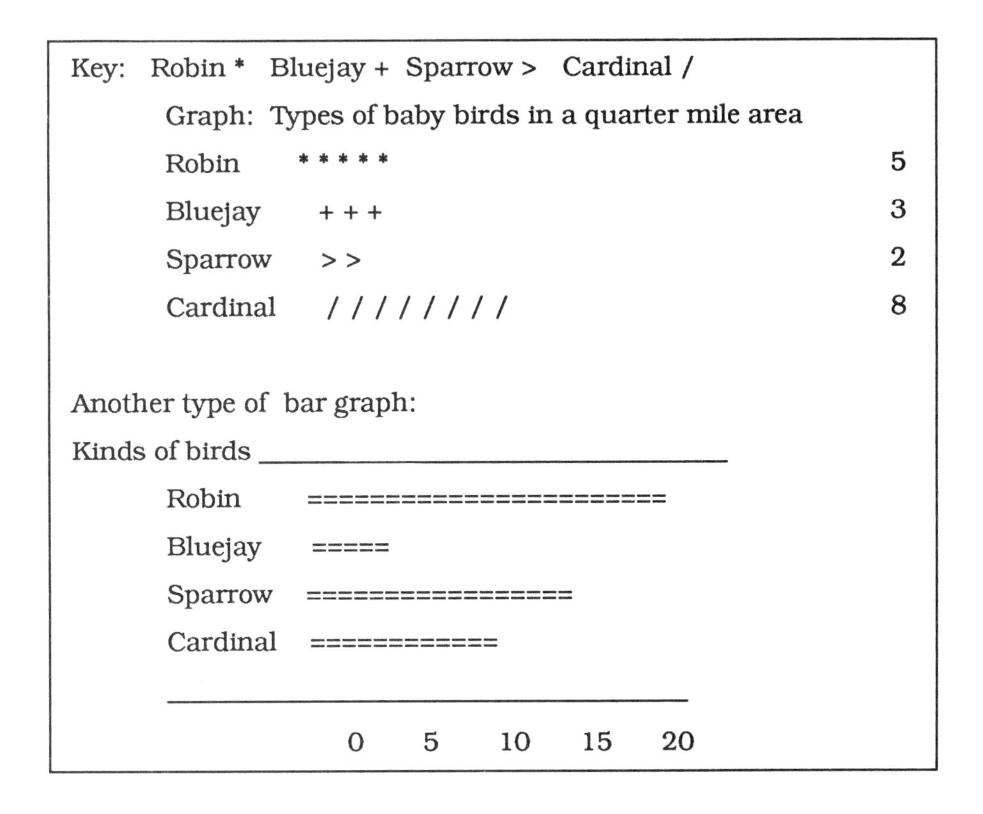

Circle graph or pie chart (shows percentages)

50% Brown hair; 25% Blond hair; 25% Red hair

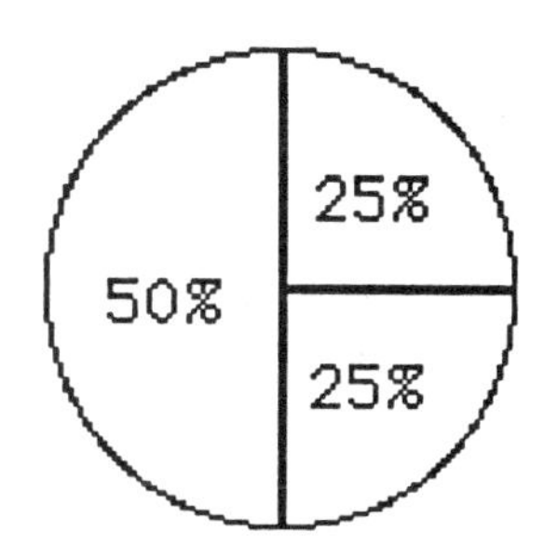

Report Reservoir

Writing a report gives you the opportunity to describe your independent project in detail, including:

A. **Front Matter**

1. Cover page with your project title, your name, date, school and grade
2. Abstract
3. Table of contents
4. Your hypothesis

B. The **main** part of your paper tells what you did and why, and how this related to research already done on this topic.

C. The **conclusion** tells what your findings show.

Remember to use your notecards when writing about your ideas. Your notecards are records of all the research you've done.

If you use a statement that someone else has said, write it with quotation marks around it. For example, "Energy is . . . ," according to Thomas Edison.

Use footnotes to tell what or who your information is from. For example, "Energy is the ability to do work."[1]

Then the footnote would look like this:

1. Johnson, Alan, Science in Our World, McRand Publishing Company, New York, c. 1980, page 3.

At the end of your written report you must supply a bibliography. This tells your reader where you got your information and when it was published. List at least four or more different sources, in alphabetical order by author's last name.

For books: after the author's last name, list: first name, name of book (underlined), where it was published, publisher, and copyright date. Also include the pages you read or used for reference.

Example: Branley, Franklin M., Energy for the 21st Century. New York: T. Y. Crowell Co., 1975, pp. 20-25.

For periodicals: author's last name, then first name, name of article in quotation marks " . . . . . ," name of periodical (underlined), volume number and page, month of the issue, and year. Collections, letters, bulletins, newspaper articles and brochures follow the same format.

Example: Hand, A. J., "Windows Tight? Now maybe it's time for energy-saving insulation shades," Popular Science, v.218, March 1981, p. 107.

For interviews: interviewer's last name, first name, job title, where it took place, and date; also include the subject of the conversation.

Example: Jones, Jordan, solar homeowner and energy consultant, interview at his home on December 12, 1987 about his passive and active systems.

For encyclopedias: title of article, name of encyclopedia, where it was published, name of the publishing company, date, volume number, pages read or used.

Example: "Energy at Work," World Book Encyclopedia, Chicago: Field Enterprises Educational Corp, 1982, Volume 15, pp. 100-125.

For government documents:

Example: Prepared by Information Planning Associates, Inc.,"What is Energy?", The Department of Energy, Office of Consumer Affairs, Washington, D.C., 1980.

No author listed:

Example: Scholastic Magazine, Senior Scholastic, Special Energy Issue, January 12, 1977.

Rest Area

What Is an Abstract?

An *abstract* is a brief summary of your project or study.

How to write your abstract:

In one page or less include

- your name, age, school, address, and grade in school
- the title of your project
- 2 or 3 paragraphs about why you did the project (what you wanted to accomplish by doing this study, what steps you took, what the final results were)

Write your abstract after you have completed the project and written your research paper.

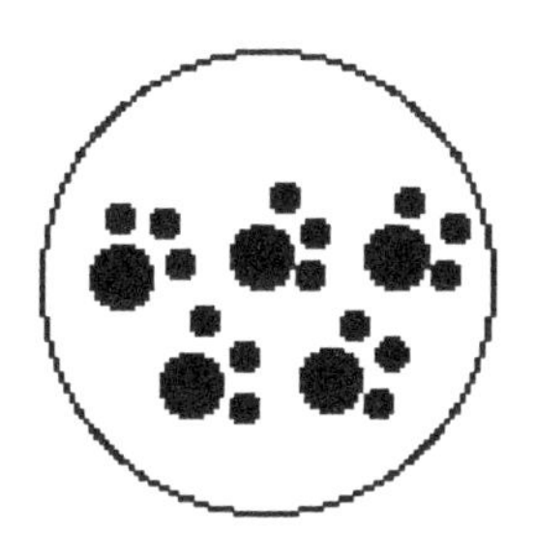

Display Trail

Now is the time to make your display. First plan **how** to do it.

Design a backboard or poster; this will be the background for your presentation. Use a sturdy material such as foam-core board or heavy corrugated cardboard. If you hinge three pieces together it will stand up by itself; you could also tape a triangle of cardboard to the back of your display to serve as an easel to support it. Cover it with white or brightly colored paper; or use fabric which can be stapled tightly around back where no one will see.

Now you have a surface on which to mount, tack or pin all the elements of your science project or independent study. Students should design their own displays, using materials from their project: photos, charts, drawings, and diagrams, enlarged so the viewer can see it easily. Consider displaying the title of your project with large, cut-out letters.

Many Science Fairs and other judged competitions have rules about how large your display may be. Typical maximum sizes are: height, no higher than 9 feet; width, no wider than 4 feet; depth, no deeper than 2 1/2 feet (distance from front to back). Be sure to check your Science Fair criteria for size restrictions.

Have your written report in a folder by your poster. Type it on a typewriter or computer, or write it very neatly. Consider making a folder with a cover designed to interest the reader in your project.

CONGRATULATIONS!

You have reached your destination.

Present your project at a Science Fair, or to an audience who will appreciate your work, and help your teacher evaluate it.

Certificate

Awarded to:

Name ______________________________

for ______________________________

____

Teacher ______________________________

Date ____________________

Name ______________________________ Date ________________

Project Title ____

SELF-EVALUATION

On a scale from 1-5, with five being highest and one being the lowest, rate the following:

	Lowest				Highest
A. I spent the right amount of time on my project.	**1**	**2**	**3**	**4**	**5**
B. My project was well planned and organized.	**1**	**2**	**3**	**4**	**5**
C. I understood my topic and what I was trying to test or solve.	**1**	**2**	**3**	**4**	**5**
D. My project answered my original question or hypothesis.	**1**	**2**	**3**	**4**	**5**
E. I used a variety of sources of information about my topic.	**1**	**2**	**3**	**4**	**5**
F. I learned a lot of new information and ideas.	**1**	**2**	**3**	**4**	**5**
G. My product showed creativity and solved problems.	**1**	**2**	**3**	**4**	**5**

continued on next page

ANSWER THESE QUESTIONS:

1. My method of research was ____

____

2. The next time I do a research project I will ______________________________

____

____

3. The best thing about this project was____________________________________

____

____

4. I could have improved my project by ___________________________________

____

____

5. What I learned from this experience was _________________________________

____

____

6. How can a future study or project be created from my research? ____________

____

____

APPENDIX

Examples of science projects and independent studies, pp. 82-93

Appendix A: CREATIVE THINKING STRATEGIES

Creative Thinking Abilities

1. **FLUENCY:** the ability to think of many ideas or responses.

2. **FLEXIBILITY:** the ability to think of different categories of thought or different ways of looking at an idea.

3. **ORIGINALITY:** the ability to think of an idea that is unique or unusual; an idea that no one else in the group may think of.

4. **ELABORATION:** the ability to add lots of detail to an idea.

(Torrance, 1966)

Creative Problem Solving

Fact-finding: Gathering all available data about a situation in order to define a problem. Ask questions such as Who, What, Where, When, Why,

Problem-finding: Analyzing subproblems and component parts of a problem in order to define the problem. Ask the question "In what ways might I . . .?"

Idea-finding: Generating ideas and uses of ideas that will lead to solutions.

Use **SCAMPER** SKILLS to think up possible ideas:

Substitute
Combine
Adapt
Modify, **M**inify, **M**agnify
Put to other uses
Eliminate
Rearrange, **R**everse

From SCAMPER, by Bob Eberle,
published by DOK Publishers, East Aurora, N.Y.
Used with permission.

Solution-finding: Evaluating, against certain criteria, the best ideas to solve a problem.

Acceptance-finding: Developing a plan of action and deciding how best to implement the best solution.

Developed by Sidney J. Parnes

Bloom's Taxonomy of Cognitive Thinking Skills

Students should be aware of the various thinking models and encouraged to use them in writing their objectives. The following are some that are commonly used:

1. **KNOWLEDGE:** The ability to learn facts; to remember or recall previously learned materials, ideas or principles.

2. **COMPREHENSION:** The ability to understand ideas or interpret information previously learned.

3. **APPLICATION:** The ability to use learned material in new and concrete situations.

4. **ANALYSIS:** The ability to break down material into parts and see relationships. This includes classifying, distinguishing and analyzing the parts.

5. **SYNTHESIS:** The ability to put together parts or ideas to form a new whole. This involves designing a plan (such as a scientific research plan), or formulating new patterns or structures; hypothesizing and composing.

6. **EVALUATION:** The ability to judge the value of material for a purpose and support this judgment with relevant criteria (such as a research report).

Reference: Bloom, B. (1956). **Taxonomy of Educational Objectives, Cognitive Domain,** printed with permission of Longman, Inc. publishers.

BLOOM'S HIGHER THINKING VERBS

Knowledge & Comprehension:		Application:	Analysis:	Synthesis:	Evaluation:
define	translate	apply	analyze	design	interpret
recognize	interpret	solve	associate	redesign	judge
recall	predict	experiment	take away	combine	decide
identify	explain	show	put together	add to	justify
label	describe	construct	combine	compose	solve
understand	summarize	interview	divide	hypothesize	infer
examine	demonstrate	sketch	isolate	construct	verify
show	discover	paint	order	reconstruct	conclude
collect	match	list	separate	translate	evaluate
observe	listen	report	distinguish	imagine	debate
locate	ask	stimulate	dissect	predict	discuss
		record	connect	assume	editorialize
		manipulate	relate	extend	recommend
		teach	differentiate	regroup	
			classify	restate	
			arrange	formulate	
			group	substitute	
			interpret	modify	
			organize	minimize	
			categorize	alter	
			take-apart	connect	
			compare/ contrast	create	

CONTENT — PROCESS — PRODUCT — RESEARCH

CONTENT

Using content related to broad-based themes, issues and problems

Promoting self-directed studies that are in-depth investigations based on student's interests

Integrating content with multiple disciplines

PROCESS

Learning fundamental skills of a discipline

Mastering complex, abstract, higher-thinking skills

Developing creative potential

Applying problem-solving procedures to a variety of problems

Developing independence

PRODUCT

Using new and varied formats for developing products

Using new materials to develop products

Using new tecniques to produce products

Expressing knowledge and sharing products with new and differentiated audiences

RESEARCH SKILLS AND METHODOLOGY

Mastering independent study and skills

Utilizing the research process

As students learn to direct their own research, gently motivate them to constantly upgrade the level of their plan.

adapted from PROJECT THINK,
Curriculum Resource Notebook,
School District No. 12, Adams County,
Northglenn, CO 80233

Appendix C

GLOSSARY

Abstract: A brief description of a report which collects the essential elements of a long paper; a summary of the project.

Analyze: To separate or break up a whole into its parts according to some plan or reason. Opposite of *synthesis*. *Structural analysis* is performed in random order. *Operational analysis* is performed in sequential steps.

Brainstorming: A group or individual method for generating solution paths for problems. The goal is to produce multiple possible solutions.

Build hypothesis: To construct tentative assumptions based on observed effect.

Categorize: To arrange items in such a way that each possesses the particular properties, based on predetermined criteria, required to belong to a specific group.

Cause/effect: A condition or event (cause) that makes something happen; the result (effect) or outcome created by the previous condition or event.

Classify: To sort into clusters, objects, events, or people according to their common elements, factors, or characteristics. Includes giving that cluster a label that communicates those essential characteristics.

Compare and contrast: To examine objects in order to note attributes that make them similar or different. To contrast is to set objects in opposition to each other or to compare them by emphasizing their differences.

Conclusion: An inferential belief that is derived from premises.

Control Group: Those factors which when run in the experiment give a predictable result. Often one factor is varied, then the results compared to the results of the control group.

Data: The actual measurements in an experiment.

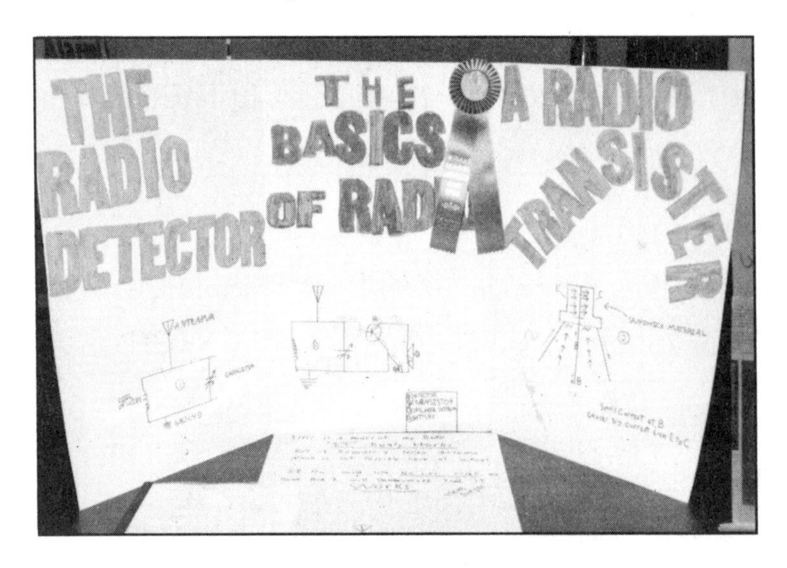

Deduce (deductive reasoning): To infer from what precedes; to lead or draw to a conclusion; to derive the unknown from the known. The opposite of *induce* (inductive reasoning).

Discriminate between definition and example: To identify a word, phrase, or term by stating its precise meaning or significance (definition) as contrasted with identifying it by giving instances of its occurrence (examples).

Discriminate between fact and opinion: To differentiate between statements generally accepted as true and those based on personal or unsubstantiated assumptions.

Discriminate between real and fanciful: To distinguish between that which is true or that which is illusory, fictitious or imaginative.

Display: To spread out for viewing; a way to present the results of the project and data collected through charts, tables, graphs, logs, drawings, diagrams, pictures and reports.

Elaborate: To expand on concepts or ideas; to give an object greater detail.

Estimate: To form a judgment about the worth, quantity, or significance of something—the implication being that the judgment formed is based on rough calculations.

Evaluation: To make an examination or judgment based upon a set of internal or external criteria.

Fact: A statement that can be proven or verified; information presented as having objective reality.

Generalization: A rule, principle, or formula that governs or explains any number of related situations.

Hypothesis: A tentative proposition or relationship assumed in order to draw out its logical or empirical consequences. An "if, then" statement that serves as a basis for testing through experimentation or gathering facts.

Hypothesize: To construct a hypothesis.

Identical: Sharing all attributes.

Induce (inductive reasoning): To combine one or more assumptions or hypotheses with available information to reach a tentative conclusion. Reaching a rule, conclusion, or principle by inference from particular facts. Opposite of *deduce*, (deductive reasoning).

Infer: To arrive at a conclusion that evidence, facts or admissions point toward but do not absolutely establish; to draw tentative conclusions from incomplete data. Inferring is the result of making an evaluation or judgment in the absence of one or more relevant facts. Inference requires supposition and leads to prediction.

Inquiry: Seeking information about a problem or condition.

Interview: A meeting with someone to obtain information about a particular subject or topic.

Irrelevant information: Data that are not useful in solving a problem or answering a question. May be a distraction.

Judgment: The process of forming an opinion or evaluation based upon value.

Knowledge: The condition of having information or of being learned.

Microfiche: Small files of lists of books, articles, and virtually all written materials usually found in libraries.

Notetaking: The act of recording observations that usually have some possible significance to an experiment or independent study.

Observe: To use the senses to gather information; to notice qualities, quantities, texture, color, form, number, position, direction, and so on.

Opinion: A personal belief, judgment, or appraisal regarding a particular matter.

Order: To arrange objects, conditions, events, or ideas according to an established scheme or criterion or to identify the scheme by which they have been arranged.

Periodicals: Magazines published regularly, such as Time Magazine or Science News.

Predict: To formulate possible consequences of a particular event or series of experiences.

Premise: A statement that allows the inference of logical conclusions.

Prioritize: To rank objects, ideas, persons, conditions, or events by importance or personal preference.

Problem: A question or situation that requires investigation and research that leads to some action.

Problem solve: To define or describe a problem, determine the desired outcome, select possible solutions, choose strategies, test trial solutions, evaluate the outcome and revise these steps where necessary.

Question: To formulate relevant inquiries so as to evaluate a situation, guide hypotheses, verify information, seek logical evidence, clarify, and so on.

Questionnaire: A set list of questions often used in a survey to gather information.

Reasoning: In two forms, deductive and inductive. *Deductive*: use knowledge of two or more premises to infer if a conclusion is valid. *Inductive*: collect observations and formulate hypotheses based upon them.

Relevant information: Data useful in solving a problem or answering a question.

Reservoir: A place where anything is collected and stored; report reservoir is a place where information about the project report is housed.

Rules: The principles or formulae that underlie or govern some problems or relationships.

Search: To delve into or over, carefully and thoroughly, in an effort to find something.

Solution: An answer or means of answering a problem.

Summarize: To present the substance of a complex idea in a more condensed or concise form.

Survey: To look over or examine with reference to condition, situation or value; a study of a specific area or number of units (such as people) with respect to a specific condition with the objective of drawing conclusions about the larger area.

Synthesize: To unite parts into a whole; to conclude; to move from principle to application; to reason deductively from simple elements to a complex whole.

Technology: The science of the application of knowledge to practical purposes; the use of computers, data bases, video equipment, lasers, cameras and other equipment to gather information and solve problems.

Test generalizations: To determine whether or not declarations, conclusions, or systematically organized bodies of knowledge (prepared by others) are justified and acceptable on the basis of accuracy and relationship to relevant data.

Variable Factor: An element, circumstance, or influence that contributes to the production of a result.

Webbing: An organizational process by which the central idea is expanded upon by branching off of it using lines connecting to circled sub-ideas.

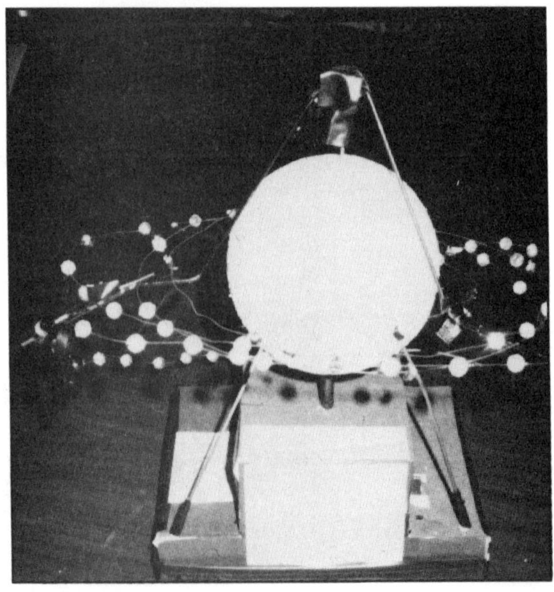

Appendix D

BIBLIOGRAPHY

Beller, Joel, **So You Want to Do a Science Project!** New York: Arco Publishing, Inc., 1982.

Bloom, Benjamin S., **Taxonomy of Educational Objectives, Handbook I: Cognitive Domain.** New York: David McKay Company, Inc., 1956.

Campbell D.T. and Stanley, Julian C. **Experimental and Quasi-Experimental Designs for Research.** Chicago: Rand McNally & Co.

Clark, B., **Growing Up Gifted.** Columbus, OH: Charles Merrill, 1983

Cory, Walter A. , Jr., "You Can't Write!", *Science Teacher Magazine*, September, 1982.

Darrow, Helen Fisher, **Independent Activities for Creative Learning.** New York and London: Teachers College Press, Teachers College, Columbia University, 1986.

Doherty, Edith J. S., and Evans, Louise C., **How to Develop Your Own Curriculum Units.** Connecticut: Synergetics, 1984.

Doherty, Edith J. S., and Evans, Louise C., **Self-Starter Kit for Independent Study.** Connecticut: Synergetics, 1980.

Draze, Diane, Blueprints, **A Guide for Independent Study Projects.** California: Dandy Lion Publications, 1986.

Erickson, Mary, "Learning to Read Educational Research,"*Phi Delta Kappan*, Dec. 1982.

Feldhusen, John F., and Treffinger, Donald J., **Creative Thinking and Problem Solving in Gifted Education.** Iowa: Kendall/Hunt Publishing Co., 1980.

Gowan, John Curtis, ed., Khatena, Joe, and Torrance, E. Paul , **Educating the Ablest,** a book of readings on the education of gifted children. Illinois: F. E. Peacock Publishers, Inc., 1979.

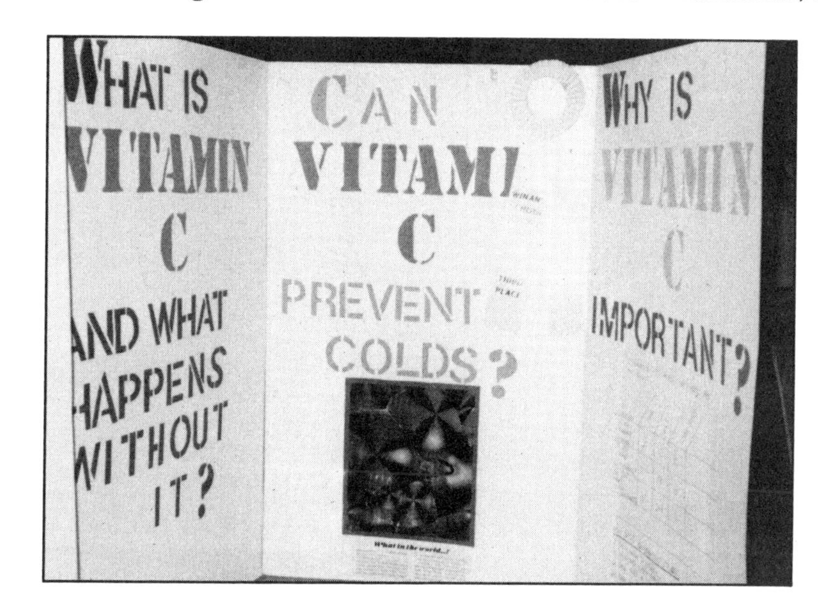

Haley, Mary Ann, Parnam, C., Purdy, P., Tamlyn, B. Smith, and Sturtevant, S. **Independent Research, Teacher's Guide and Student Notebook.** Massachusetts: IRSN, 1983.

Homeratha, Linda, and Treffinger, Don, **Independent Study Folder.** New York: D.O.K. Publishers, 1979.

Knowles, Malcolm, **Self-Directed Learning.** New York: Cambridge, The Adult Education Company, 1975.

Leiker, Dr. Mary, **An Affordable Gifted Program That Works.** Denver, CO: Coronado Hills, Third printing, 1982.

McDonnel, Rebecca C., **Gifted Independent Study.** Sewell, NJ: EIRC, 1981.

McKenna, Alexis, **Doodling Your Way to Better Recall, or Mapping Your Way to Memory.** Tucson, AZ: Zephyr Press, 1979.

McKisson, Micki, **CHRYSALIS, Nurturing Creative and Independent Thought in Children.** Tucson, AZ: Zephyr Press, 1983

National Science Teachers Association, **Science Fairs and Projects.** Washington, DC: NSTA, 1985.

Noller, Ruth, and Treffinger, Donald J., Houseman, Elwood D., **It's A Gas To Be Gifted,** New York: D.O.K. Publishers, Inc., 1979.

Perry, Phyllis J., and Hoback, John R., **A Guide to Independent Research,** Alabama: GTC, Inc., 1986.

Polette, Nancy, **The Research Book For Gifted Programs K-8,** Missouri: Book Lures Inc., 1984.

Rand McNally Road Atlas. Chicago, IL: Rand McNally & Co., 1986.

Renzulli, J. S., **The Enrichment Triad Model: A guide for Developing Defensible Programs for the GIfted and Talented,** Mansfield Center, CT: Creative Learning Press, Inc., 1977.

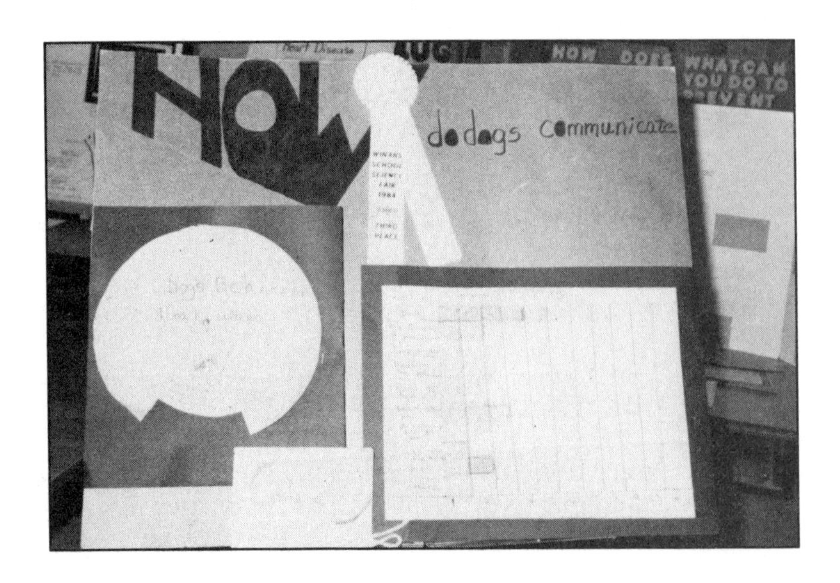

Roets, Lois Schelle, Ed.D., **Survey and Public Opinion Research,** An Instructional Program. New Sharon, IA: Leadership Publishers, 1987.

Rogers, C., **Freedom to Learn for the 80's.** Columbus, Ohio: Charles Merrill, 1983

Saul, Wendy, **Science Fare, An Illustrated Guide and Catalog of Toys, Books and Activities for Kids.** New York: Harper and Row, 1986.

Science Service, **Thousands of Science Projects,** Classified Titles of Exhibits Shown at Science Fairs and/or Produced as Projects for the Westinghouse Science Talent Search, Washington, DC: Science Service, 1987.

Stanish, Bob, **Hippogriff Feathers.** Carthage, IL, Good Apple, Inc., 1981.

Stwertka, Eve and Albert, **Make It Graphic!, Drawing Graphs for Science and Social Studies Projects.** New York: Julian Messner.

Van Deman, Barry A., and McDonald, Ed, **Nuts and Bolts: A Matter of Fact Guide to Science Fair Projects.** Illinois: The Science Man Press, 1980.

Van Tassel-Baska, Joyce, Ed., **A Handbook on Curriculum for the Gifted.** Illinois: Northwestern University, A Publication of the Center for Talent Development.

Villalpando, Eleanor, and Kolbe, Kathy, **Independent Projects.** Phoenix, AZ: Think Ink Publications, 1979.

Wishau, Jan, and illustrated by Jean Thorley, **Investigator: A Guide for Independent Study Projects**. California: Dandy Lion Publications, 1985.

Wurman, Richard Saul, editor, **Yellow Pages of Learning Resources.** Pennsylvania: Resources Directory, Group for Environmental Education, Inc., Seventh printing, 1979.

Outstanding Science Trade Books For Children, a list compiled yearly by the National Science Teachers Association and Children's Book Council Joint Committee. Children's Book Council.

RESOURCES

Bulletin Board Planner. Students research, design and create classroom or library bulletin boards on their own or in small groups. Useful in sharing results of independent study. Zephyr Press, Tucson, AZ.

Interest-A-Lyzer. An in-depth questionnaire for the individual student. Encourages targeting on special interest before selecting a research project. Creative Learning Press, Mansfield, CT.

Learning Styles Inventory: A Measure of Student Preference for Instructional Techniques. Renzulli, J. S., and Smith, L., Mansfield Center, CT: Creative Learning Press, 1977.

Media Magic. A Filmstrip-Making "Center-in-a-box." Simple-to-prepare materials for making slides and filmstrips. Zephyr Press, Tucson, AZ.

Science Fair Project Index. Prepared by the Science and Technology Division of Akron-Summit County Public Library. Assists students in developing ideas for science fair projects. Scarecrow Press, Inc., Metuchen, NJ.